USAGE OF RADIORECEPTOR ASSAY TO STUDY THE MECHANISM OF LUTEINIZING HORMONE IN SOME OF OVARIAN TUMORS

PROF. DR.SAMI A. AL-MUDHAFFAR
DR. NADA ABDUL-KAREEM ABU AL-TEMAN

Chapter 1

Introduction
and
Literature Survey

1.1 The Ovaries

1.1.1 Anatomical Consideration

The female reproductive organs consist of *ovaries, fallopian tubes, uterus, vagina, external genitalia* and *mammary glands*. The *ovaries* are paired pelvic organs that lie on either side of the uterus close to the lateral pelvic wall, behind the broad ligament and anterior to the rectum, below the fimbriated ends of the two fallopian tubes[1-3].

Adult ovaries are ovoid, measure approximately 3-5 cm in greatest dimension, and weigh 5-8 gm. However, size and weight vary considerably depending on age and content of follicular derivates. They have grayish pink appearance and an irregular surface[4, 5].

The ovarian arteries, veins and nerves traverse the suspensory ligament and enter the ovary through the mesovarium and thence through long channals to terminate at the level of the kidneys, the major lymphatic drainage of the ovary is therefore cephalad towards the para-aortic nodes. A layer of visceral peritoneum covers the surface of the ovary[4,6,7].

1.1.2 Histology

The ovary (Figure 1-1) is comprised of three distinct regions[8, 9]:

1. *An outer cortex*   : containing the ovarian follicles.
2. *A central medulla :* consisting of ovarian stroma.
3. *An inner hilum*   : around the area of attachment of the ovary to the mesovarium.

Ovarian cortex contains the ovarian follicles at different stages of maturation (Primary, Secondary, Tertiary, Graafian and atretic) together with corpora lutea and corpora albicantia for those that have reached full maturation[10].

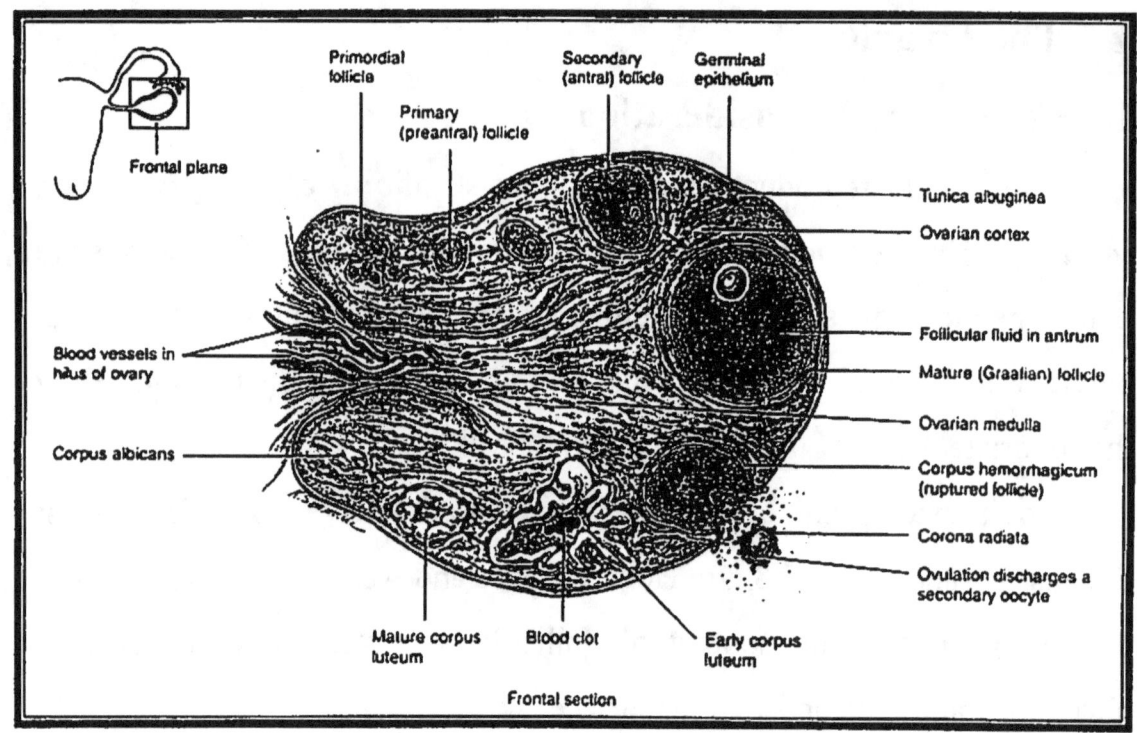

Figure (1-1): Histology of the ovary[9]

The ovary is invested with mesothelial covering, like other organs of the abdominal cavity. These modified peritoneal cells which cover the surface of the ovary are the most common source of ovarian neoplasm, and these epithelial tumors account for 90% of ovarian carcinomas[4, 11, 12].

Ovarian stroma consists of three specific cell types, contractil cells, connective tissue cells, which function to give structural support and interstitial cells[13].

The hilum contains specific type of intersitial cells known as the hilus cell. Normal hilus cells have been shown to synthesize and secrete testosteron in response to Luteinizing Hormone (LH)[14, 15].

1.1.3 Ovarian Physiology

1.1.3.1 Ovarian Function

The ovaries fulfill two important physiological functions, *reproduction and control of secondary sex characteristics*. These hormonal functions reside in the ovarian follicle, which makes fertility possible by releasing the ovum and regulating the menstraul cycle through the secretion of estrogen and progesterone hormones. In addition, estrogens stimulate the development of female secondary sex characteristics[5].

The Ovaries also function during fetal life because of stimulation by another gonadotrophic hormone, *human chorionic gonadotrophin (hCG)*, secreted by the placenta, but within a few weeks after birth this stimulus is lost, and the ovaries become almost dormant until the prepubertal period[1].

1.1.3.2 Menstrual Cycle

This term refers to the series of changes that occur in sexually mature, non-pregnant females and that culminate in *menses*. Menses is a period of mild hemorrhage during which part of the endometrial lining of the uterus is slaughed and expelled from the uterus. Typically the menstrual cycle is about 28 days long, although it may be as short as 18 days or as long as 40 days[6].

A series of cyclic changes in gonadal steroid hormones production characterizes adult ovarian function (Figure 1-2). This monthly steroid hormones profile results from cyclic changes in pituitary gonadotrophins (Gn). These, in turn, reflect changing pituitary sensitivity to gonadotrophin releasing hormone (GnRH)[16]. The first day of the menstrual bleeding (menses) is considered to be day 1 of the menstrual cycle. Menses typically lasts 4 or 5 days. Ovulation, occurs on about day 14 of the menstrual cycle[6].

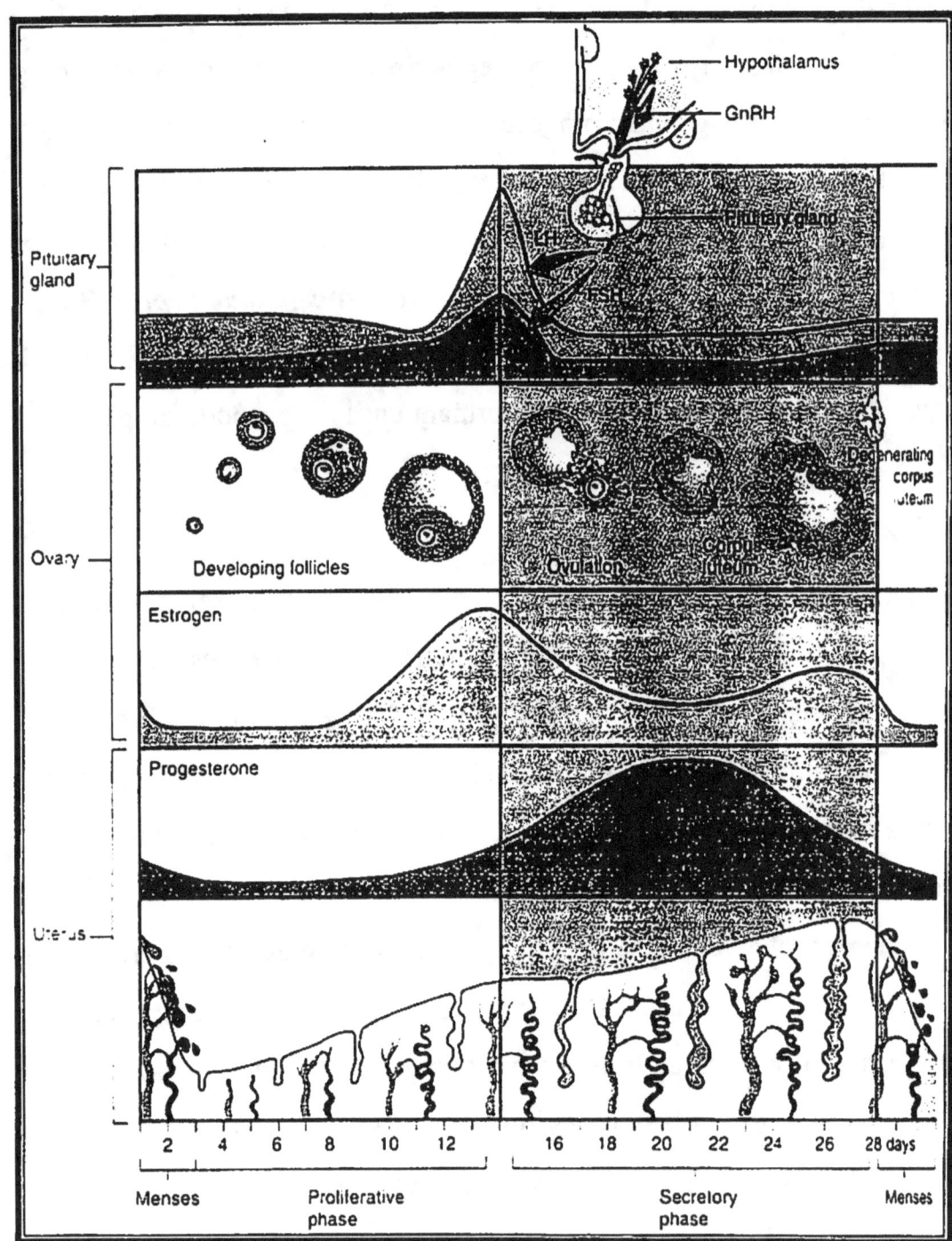

Figure (1-2): Menstrual Cycle [6]

1.1.3.3 Follicular Development and Ovulation

Primordial follicle composed of primary oocyte (the female germ cell) surrounded by a single layer of flat cells called granulosa cells, from birth to puberty the number of primordial follicles decline from 2 millions to 300,000 – 400,000[1].

The initial change that results in puberty appears to be maturation of the hypothalamus, the hypothalamus and anterior pituitary secrete large amounts of *GnRH, LH* and Follicle Stimulating Hormone *(FSH)*. Every 28 days, FSH and LH stimulate primordial follicles to develop into mature follicle or graafine follicle. Although several follicles begin to develop during each ovulatory cycle, only one or two complete the entire developmental process and rupture to release a mature ovum, while other follicles degenerate through a process called *atresia*[6,8].

As the follicles ripens, ovarian estrogen is produced by the granulosa cells. Estrogen begins to stimulate the development of the endometrium in the uterus. The sustained increase of estrogen secreted by the developing follicles stimulate GnRH secretion from the hypothalamus and GnRH triggers LH and FSH secretion from the anterior pituitary gland. This positive feed back loop produces a series of large surges of LH and FSH. Estrogen in concert with FSH from the mature follicle causes an increase in LH receptors on granulosa cells. LH acts to augment progesterone secretion by granulosa cells, which stimulates FSH release at midcycle. Following ovulation, the number of LH receptors in the lutein cells increases and the FSH receptor numbers and FSH responses decreases. Ovulation, a rupture of the follicle and releasing the ovum, occurs in response to large increase in LH levels and normally occur on about day (14) of the menstrual cycle[3,8,17].

The ovum normally is then picked up and transported through the fallopian tube toward the uterus. This large increase in LH is also responsible for the development of corpus luteum (yellow body), which develops from a mature graafian follicles after ovulation [8,18]. The corpus luteum is the principle source of progesterone, and the latter acts on the uterus causing the cells of the endometrium to become larger and to secrete a small amount of fluid. Together, progesterone and estrogen inhibit LH and FSH secretion, thus LH and FSH levels decline to low levels after ovulation[4,19] (figure 1-2).

If *fertilization* does not take place, the corpus luteum atrophies and is replaced by white scar tissue (corpus albicans), the progesterone support of the endometrium is withdrawn and menstruation occurs[3]. If fertilization has occurred the zygote has undergone several cell divisions to produce a collection of cells called blastocyst. The endometrium is prepared to receive the blastocyst which becomes implanted in the endometrium[6].

Human chorionic gonadotrophins is produced by the trophoplastic cells within the implanted blastocyst and the primary function of hCG is to support the corpus luteum which remains functional for 3 months until the placenta produces amounts of progesterone sufficient to support the pregnancy.[20,21]

1.1.3.4 Ovarian Hormones and Female Secondary Sex Characteristics

The ovaries produce *estrogen*, *progesterone* and *androgens* (figure 1-3).The changes associated with puberty are primarily the result of the elevated rate of estrogen and progesterone secretion by the ovaries[3].

The first signs of puberty can appear as early as age 8 in girls, and the process is largely completed by age 16. puberty in females is marked by the 1st episode of menstrual bleeding which is called *menarch*. During puberty, the vagina, uterus, uterin tubes and external genitalia begin to enlarge. Fat is deposited in the breasts and around the hips causing them to enlarge and assume an adult form. In addition, pubic and axillary hair grows. Development of sexual drive is also associated with puperty[6].

Estrogens are necessary for the normal physical maturation of the females. They promote development of the above secondary sexual characterestics of women and promote uterine growth, thickening of vaginal mucosa, thinning of the cervical mucus, and development of ductal system of the breast[3.22].

Progesterone is required for the implantation of the fertilized ovum and maintenance of pergnancy. Progesterone also inhibits uterin contractions, increases the viscosity of cervical mucus, promotes glandular development of the breast, and increases basal body temperature[22,23].

Androgen has an important anabolic effect on protein metabolism and influence sexual libido in the female[23].

Figure (1-3): Chemical formulas of the principle female hormones[1, 23]

1.1.3.5 Menopause

When a woman is 40–50 years old, the menstrual cycles become less regular, this is called the *female climacteric*, eventually the cycles stop completely which is the *menopause*[8, 24].

The major cause of menopause is age- related changes in the ovary. Successive cycles of ovulation and atresia deplete the ovarian content of its follicles. The median age of menopause is 50-51 year. Ovarian carcinoma continue to be a disease that affects predominantly postmenopausal women[17, 25, 26].

Postmenopausal ovary is reduced in size, weighing less than 2.5 g, and is wrinkled or prune – like in appearance. Furthermore, the loss of ova and follicles results in a reduction of the cortical area, the most striking changes occur as the stroma becomes hyperplastic and dominates the ovary[27,28].

During menopause, the continuos loss of ovarian follicles results in a decrease in estrogen secretion, serum FSH and LH increase markedly during the menopause as the result of decreased feed back inhibition by ovarian steroids and possibly inhibin, and the normal pulsatile pattern of gonadotrophin secretion is exaggerated[5].

During climacteric, some women experience, irritability, Fatigue, anxiety, temporary decrease in libido and occasionally severe emotional disturbances, decreased estrogen formation also triggers vasomotor instability, atrophy of the uro-genital epithelium, a reduction in size of reproductive organs and breast, an increased risk of cardiovascular disease, and osteoporosis. Estrogen therapy of these symptoms has a potential side effect which is slightly increased possibility of the development of breast and uterin cancer in some women[6,8].

1.2 Ovarian Diseases

1.2.1 Premature Ovarian Failure [4]

   a.   Developmental malformation (ovarian dysgenesis)
   b.   Ovarian destruction / ablation (surgical, radiation – induced, drug – induced).
   c.   Premature follicular depletion (true premature menopause).
   d.   Gonadotrophin resistant ovary (insensitive follicles to gonadotrophins).
   e.   Auto – immune oöphoritis.

1.2.2 Inflammation and Infection[12]

   a. Acute and chronic pelvic inflamatory diseases.
   b. Tuberculosis of the genital tract.
   c. Sexually transmitted diseases (e.g. Gonorrhea).

1.2.3 Non- Neoplastic Functional Cysts[29,30]

a. follicular cyst.

b. Corpus luteum cyst.

c. Theca lutein and granulosa lutein cysts.

d. Polycystic ovarian disease.

e. Endometriomatous cysts.

1.2.4 Ovarian Neoplasms

The ovary gives rise to a wider variety of tumors than any other organ in the body. Natural history and response to treatment vary considerably from one tumor group to another, accordingly, accurate histological diagnosis is often a critical factor[31].

The tumors may be solid, cystic or a mixture of both, and may be benign, malignant or in a borderline state. Malignant neoplasm of the ovaries can be divided into three categories: *epithelial tumors*, *germ cell tumors* and *gonadal stromal tumors*[4,32] (table 1-1).

Most of the ovarian cancers are epithelial in origin, as germ cell tumors of the ovary represent only ~5% of the total and 65% of benign ovarian tumors are epithelial in origin. Serous epithelial tumors form about 40% of ovarian tumors and 50-70% of these are benign serous cystadenoma[31, 33].

The relationship of age to histologic type of ovarian neoplasm is confirmed. Women older than age 30 are unlikely to be diagnosed with germ cell tumors. Likewise, very few borderline carcinomas are found in women older than age 65. In addition, germ cell and borderline carcinomas are each much more likely to be found at an early stage compared with other carcinomas [26, 34].

Table (1-1): World Health Organization (WHO) Classification of Ovarian Tumors [4]

Surface epithelial stromal tumors	Serous tumors	Benign (cystadenoma) Of borderline malignancy Malignant (serous cystadenocarcinoma)	
	Mucinous tumors, endocervical-type and intestinal-type	Benign Of borderline malignancy Malignant	
	Endometrioid tumors	Benign Of borderline malignancy Malignant Epithelial–stromal	Adenosarcoma Mesodermal (müllerian) mixed tumor
	Glear cell tumors	Benign Of borderline malignancy Malignant	
	Transitional cell tumors	Brenner tumor Brenner tumor of borderline malignancy Malignant Brenner tumor Transitional cell carcinoma (non-Brenner-type)	
	Undifferentiated carcinoma		
Sex cord-stromal tumors	Granulosa-stromal cell tumors	Granulosa cell tumors Tumors of the thecoma-fibroma group	
	Sertoli-stromal cell tumors (androblastomas)		
	Sex-cord tumor with annular tubules		
	Gynandroblastoma		
	Steroid (lipid) cell tumors		
Germ cell tumors	Dysgerminoma		
	Yolk sac tumor (endodermal sinus tumor)		
	Teratoma	Immature Mature (adult) Solid Cystic (dermoid cyst)	
	Monodermal (e.g., struma ovarian, carcinoid)		
	Mixed germ cell tumors		
Gonadoblastoma			
Tumors not specific to ovary			
Unclassified			
Metastatic tumors			

1.3 Epithelial Ovarian Tumors

The most common group of ovarian neoplasm originate from the coelomic mesothelium that covers the ovary, which after neoplastic transformation seems to retain the capacity to recapitulate the epithelial components of the mullerian ducts, for example, the epithelium of serous tumors resembles that lining the fallopian tube. According to the *World Health Organization (WHO)* classification of ovarian tumors, surface epithelial – stromal tumors can be divided into *serous tumors, mucinous tumors, endometrioid tumors, clear cell tumors, transitional cell tumors* and *undifferentiated carcinoma*[4, 35] (Table 1-1).

Based upon morphology, epithelial tumors have been subdivided into *benign, borderline* and *malignant subcategories*. Evidence is lacking about whether ovarian carcinoma may go through a borderline phase during its development and whether borderline tumors always shift into invasive ovarian carcinoma [36-38].

Benign serous cystadenomas occur slightly more often than benign mucinous tumors, but in their malignant form serous cystadenocarcinomas are three to four times more common than mucinous cystadenocarcinomas. Serous and mucinous borderline tumors are seen but other types of epithelial tumors of borderline malignancy, such as the variants of Brenner and endometrioid tumors, are rare[31].

1.3.1 Incidence

Ovarian cancer is the second most common female genito-urinary cancer, and the most lethal. It is the seventh most common malignant tumors among the women in Iraq. The *Iraqi Cancer Registry* estimates that in Iraq there is a three fold increase in the incidence of this disease during the last two decades. The overall risk of developing ovarian carcinoma is approximately 1 percent [11, 39, 40]. The incidence of ovarian cancer increases with age, being greatest between ages 65 and 84. Approximately 70% of cases will occur in women who are over the age of 50 year [41,42]. Only 10% are younger than age 40 and merely 3-4% are younger than age 30 years, whereas the greatest number of cases occur in the postmenopausal age group, so, it is rare in the reproductive age group[37].

1.3.2 Etiology

While the etiology remains unknown, the most significant risk factor for ovarian cancer appears to be the *ovulatory age*. Data supporting this theory are increased risk with nulliparity, and with female received drugs that induce ovulation but have not become pregnant, and decreased risk in women using oral contraceptives [3, 43- 45].

Ovarian cancer probably arises due to *accumulation of mutations* in multiple combination of genes, mutation and over expression of p53 gene are currently the most common, occuring up to 90% of advanced epithelial cancers and correlates with aneuploidy, high grade, advanced stage and unfavorable prognosis [46-51].

The principle risk factor, however, is a *family history of ovarian cancer (hereditary cause)*, particularly when two or more first degree relatives have been effected.

From a histogenic view point, ovarian cancer may develop de novo or may arise from *pre-existing benign epithelial tumors* [52, 53].

Carcinogens in the peritoneal cavity proposed to be a propable cause, studies have shown that women who use talcum powder as part of their perineal hygiene are at increased risk. Talc is found in soap powders and deodorants, and is used in the packing of condoms and contraceptive diaphragms. Talc might then migrate through vagina to reach the ovaries[31,54,55].

It was reported that risk for ovarian cancer increased in woman previously treated with ***pelvic irradiation***[31].

Hormonal role in carcinogenesis studies have shown that gonadotrophins may play a role by indirect or paracrine means, acting as promoters or cofactors in carcinogenesis. It has been postulated that elevated gonadotrophins (Gn) levels may contribute to malignant transformation. Decreased risk in women using oral contraceptives support this postulation as oral contraceptives cause a reduction in pituitary Gn[56, 57]. Hence, high levels of gonadotrophins in menopausal women could have a potentiating influence on pre-existing pre-malignant or malignant changes in the ovary. The hormonal basis for ovarian cancer appears to differ from that of other hormone-induced cancers in that the responsible hormones (gonadotrophins) act indirectly on ovarian epithelial cells. The stimulus for cell division in the etiology of ovarian cancer is not hormonal per se; rather it follows ovulation, which is the direct result of complex hormonal changes[58,59]. Reduced risk by oral contraceptives propably because they suppress ovulation, they may act by reducing the repetitive trauma on the ovarian surface, as well as the exposure of the surface epithelium to the estrogen-rich follicular fluid, which in turn may influence mitotic proliferation, uncontrolled cell division and subsequent malignant transformation[60].

1.3.3 Pathology

Serous tumors varies from smooth- walled cystic forms (simple serous cystadenoma) to superficial papillary neoplasm without cystic cavity (superficial papilloma).They can be of any size up to 25 cm in diameter,and usually smaller than the mucinous variety [11, 31, 61].

Unilocular cystadenomas are more frequent than multilocular form, borderline ones are often multilocular. Papillary projections constitute a hallmark of serous tumors, in benign forms the papillae are small and less numerous than in malignant forms, whose papillae have proliferated so much that they almost fill the cavity, utimately perforating the capsule. Cysts contain thin serous fluid, which may be flocculated or bloody due to secondary changes in the wall. 40-50% of serous carcinoma affect both ovaries, while only 5-8% of benign forms are bilateral[2, 11, 61].

Benign serous cystadenomas are lined by a single layer of low columnar epithelium, which is sometimes ciliated and resembles tubal epithelium. Papillary projections are common, and may be branching. They are composed of well differentiated cells and qualified as benign[4.61]. Borderline serous tumors are characterized by epithelial proliferation greater than that seen in benign tumors but with absence of "destructive" invasion of the stroma, minimal invasion does not indicate true malignancy[61-63].

Serous carcinomas have an anaplastic epithelial component that invade the underlying stroma, the epithelial cells are usually several layers thick and have anaplastic nuclei with loss of polarity[4].

1.3.4 Staging[4,31,54,64]

Staging of ovarian cancer is based on the findings at the time of surgery and pathological review. Because of the clinically occult spread, surgery is mandatory.

Current data suggest that in apparent stage I and II disease cytological washings will be positive in 10-50% of cases, positive para-aortic nodes will be found in up to 20% of cases, omental secondaries in 0-4.7% of cases, diaphragmatic metastasis in 0-44% of cases. Accurate staging is of utmost importance for the patient's further therapy and for discussing prognosis. The latest collected world data of *International Federation of Gynecology and Obstetrics (FIGO)* on the distribution of the patients in the different stages show 26.1% in stage I, 15.42% in stage II, 39.1% in stage III and 16.3% in stage IV. A total of 3.1% were unstaged (Table 1-2).

Table (1-2): (FIGO) Staging for Primary Carcinoma of Ovary [31]

Stage I	Growth limited to the ovaries
Stage 1a	Growth limited to one ovary; no ascites. No tumor on the external surface; capsule intact
Stage Ib	Growth limited to both ovaries; no ascites. No tumor on external surfaces; capsules intact.
Stage Ic	Tumor either Stage Ia or Ib but with tumor on surface of one or both ovaries; or with capsule ruptured; or with ascites present containing malignant cells or with positive peritoneal washings.
Stage II	Growth involving one or both ovaries with pelvic extension.
Stage IIa	Extension and/or metastases to the uterus and/or tubes
Stage IIb	Extension to other pelvic tissues.
Stage IIc	Tumor either Stage IIa or IIB but with tumor on surface of one or both ovaries; or with capsule(s) ruptured ; or with ascites present containing malignant cells or with positive peritoneal washings.
Stage III	Tumor involving one or both ovaries with peritoneal implants outside the pelvis and/or positive retroperitonal or inguinal nodes. Superficial liver metastasis equals Stage III. Tumor is limited to the true pelvis but with histologically proved malignant extension to small bowel or omentum.
Stage IIIa	Tumor grossly limited to the true pelvis with negative nodes but with histologically confirmed microscopic seeding of abdominal peritoneal surfaces.
Stage IIIb	Tumor of one or both ovaries with histologically confirmed implants of abdominal peritoneal surfaces. none exceeding 2 cm in diameter. Nodes are negative.
Stage IIIc	Abdominal implants more than 2cm in diameter and/or positive retroperitoneal or inguinal nodes.
Stage IV	Growth involving one or both ovaries, with distant metastases. If pleural effusion is present, there must be positive cytology to allot a case to Stage IV. Parenchymal liver metastases equal Stage IV.

1.3.5 Diagnosis

1.3.5.1 Clinical Features

Ovarian tumors are symptomless in their early stages, and are only discovered during routine examination. Even when symptoms are present they are at first non-specific and many patients are initially referred to specialist other than gynecologist. Many patients had months of gastro-intestinal symptoms including belching, early satiety, abdominal fullness or duspepsia. Lower abdominal discomfort, frequency similar to the symptoms of irritable bowel syndrome and diverticulitis. Most patients with ovarian cancer have loss of appetite, ill health and sometimes weight loss, and the majority of patients with serous cystadenocarcinoma have advanced disease at the time of diagnosis[11,40,54,59,65,66].

As the disease progress, symptoms become more specific and constant, related to pain and pressure caused by enlarging tumor or ascitis which is not uncommon. Pressure symptoms include dyspnea, lower limb oedema, varicosities and hemorrhoids, deep venous thrombosis may be the presenting sign, frequency, urgency and acute urin retention may occur. Intestinal obstruction is not uncommon. In terminal stage of ovarian cancer cachexia is common[31, 54].

The acute pain caused by torsion or hemorrhage or rupture (complication of ovarian tumor) into the tumor are infrequent symptoms. The definitive diagnosis of ovarian cancer is based on the results of surgical exploration and pathological review, but certain physical signs may indicate malignancy as solid tumor, irregular surface, ascitis which is common, fixity which is usual and bilaterality, while benign tumors are usually small, smooth surface and mobile [34,54,59].

1.3.5.2 Investigation[31, 67, 68]

Abdominal and *pelvic ultrasound* is useful in establishing the size and site of tumors and show many features that may suggest malignancy, but it may not demonstrate with certainty the presence of tumor deposits of less than 2 cm in diameter, nor differentiate the benign from the malignant with a positive predictive sensitivity of more than 75%. Colour- flow doppler techniques appear to improve this sensitivity. *Computed Tomography (CT Scanning)* and *nuclear magnetic resonance imaging (MRI)* are more used, but it fail to differentiate benign disease from stage I disease and whether adhesions and distant lesions were due to malignant or unrelated pathology, however, a *CT*, or *MRI* may be useful in identifying other potential sites of origin. *Laparoscopy* can be useful if the exclusion of ovarian cancer means a laparotomy can be avoided. *Radiological studies of the upper and lower gastrointestinal tract, chest -x- ray and intravenous urography* are necessary if the ovarian tumor is thought to be metastatic.

1.3.5.3 Tumor Markers

Expression of specific antigens is useful for establishing a diagnosis and classification and providing prognostic information, monitoring the appropriate antigen titers is very useful in identification of occult metastasis, monitoring of therapeutic response and detection of asymptomatic recurrence at an early stage[68-71].A *carbohydrate antigen (CA-125)* is the first serum tumor marker test for epithelial ovarian cancer, was introduced by *Bast et al* [72]. CA-125 serum concentration has increased in approximately 50% of localized stage I, 90% of advanced disease stage (II-IV), in 80% of non mucinous epithelial cancer, it's associated with serous tumors while mucinous tumors often produce carcino-embryonic antigen (CEA). CA-125 level is also elevated in 25% of non gynaecological malignancies, in 5% of benign disease and only in 1% of healthy people[70.72-75].

The protein (p53) mutations are common (50-80 %) in ovarian cancer. The frequency of mutations increases in advanced-stage of epithelial tumors. There is a statistically significant association between p53 staining and higher-grade disease, and there is no apparent predilection for p53 abnormalities in serous versus non serous histologic subtypes[76-79].

Tumors associated glycoprotein (TAG-72) level is elevated in 50% of ovarian carcinoma cases and only in 4% of benign diseases cases with the highest level of expression in mucinous cystadenocarcinoma and its measurement may be useful as a confirmatory tumor marker for the presence of ovarian cancer in those patients with elevated CA-125 serum levels. Combined TAG- 72 and CA-125 test increases the sensitivity for the detection of primary ovarian cancer from 60% to 73% with no significant change in specificity[80-84].

1.3.6 Treatment

Since a pelvic tumor may be malignant, surgical exploration or laparoscopy is essential, in particular, ovarian tumor should always be operated on. Another decision must be taken about the extent of surgery once abdomen has been opened, and this depends on the nature of the tumor and the age of the patient[11].

- *Benign Serous Tumors*: Whenever benign disease is confirmed there is apportunity to consider conservative treatment[2, 31].

   1. *Simple serous cyst*: Treated by simple cyst enucleation with preservation of both ovaries.

   2. *Large cyst or solid component in cyst*: Treated by unilateral salpingo- oöpherectomy (in young women).

3. *In women after 45 years old*: The treatment is bilateral salpingo-oöpherectomy and hysterectomy.

- *Borderline Serous Tumors:*[54,85,86]

    1. *Stage Ia (limited to one ovary)*: Treated by unilateral salpingo-oöpherectomy. This surgical resection affords anearly 100% survival.

    2. *Stage III and stage IV*: No definite treatment plan has been established. Surgical resection (aggressive) may offer best chance for prolonged palliation and long term survival. Responses to chemotherapy or radiation therapy are not well documented, but they are given in proved epithelial ovarian cancer (transformed borderline).

- *Serous Cystadenocarcinoma*: Clinical management of ovarian cancer divided into surgery and adjunctive therapy. Two goals of surgery are (1) *accurate surgical staging* (2) *tumor resection*. Adjunctive therapy is required for most patients to improve their chance of survival and provide effective palliation, its type depends on the disease stage and nature of residual disease[54,87].

    1. *Early stage Ia*: Conservative unilateral salpingo oöpherectomy is adequate, they do not benefit from adjunctive therapy. It appears that ovarian preservation (and woman's fertility) is safe and reasonable in women of reproductive age[54,88].

    2. *Stage I, II*: Total abdominal hysterectomy with bilateral salpingo-oöpherectomy, pelvic and para-aortic lymphodenectomy. Surgery alone give approximately 50-60% five years survival, with addition of adjunctive therapy survival becomes 85-90% [54,89].

    3. *Advanced cancer stage III, IV*: There is often extensive intra-abdominal extrapelvic disease and complete removal may not be possible, so, removing the maximum amount of tumor offers the best

chance of response to chemotherapy in the postoperative period. This is known as cytoreductive or debulking surgery and the optimal debulking is that in which the diameter of residual tumor nodules is less than 1cm. Cytoreductive surgery includes omentectomy, intestinal resection, total abdominal hysterectomy with bilateral salpingo-oöpherectomy and recto-sigmoid resection. Extensive surgery is often insufficient to eleminate the intraabdominal tumor and response to chemotherapy is only partial in many of these patients[54,90,91].

Post- operative adjunctive therapy choices: [54, 92, 93]

- *Whole abdomen radiation therapy*: It is associated with high morbidity and injury to liver, kidney and small bowel, if used, to be only of value in stage III patients with minimal or no gross residual disease.

- *Intra peritoneal chromic phosphate (^{32}P)*: Because ^{32}P does not effectively irradiate the retroperitoneal lymphnodes, its use should be reserved for stage III patients who had lymphadenectomy and showed no evidence of metastasis.

Chemotherapy

Ovarian cancer is one of the most responsive human malignancies to cytotoxic chemotherapy, combination therapy is shown to achieve a higher response rate than do treatment with single agent[31, 94] A combination of paclitaxel and cisplatin appears to be the most effective regimen, it increases the median survival from 24.4 to 37.5 months (54%), over the usual cure in the treatment of advanced epithelial cancer, and appears to be cost-effective first line treatment for advanced ovarian carcinoma. A synergistic activity of paclitaxal combind with ifosfamide, has been reported in ovarian cell lines[95 - 99].

Although the ovarian cancer is chemosensitive, cures with available agents are most unusual, but the patients live longer in remission[111]. Cytotoxic drugs are usually given intravenously, and other known routs, an alternative approach for delivery of these drugs is intraperitoneal administration, in which higher doses may be delivered to the tumor and the systemic toxicity is minimized, this technique is effective in advanced disease with minimal residual disease[54].

Second-look procedure

In some institution, "Second-look" operations have become a standard procedure for patients with ovarian carcinomas treated with surgery and chemotherapy who are clinically free of disease, to determine prognosis and decide whether chemotherapy should be continued, discontinued or changed[100,101]. The second–look operation includes direct visual inspection of the abdominal cavity, cytologic examination of peritoneal washings and multiple biopsies of peritoneum, omentum and lymphnodes. Residual tumor has been found pathologically in over 40% of the patients in most series[102-104].

Studies on hormonal role in ovarian cancer treatment

The responses to hormonal therapy in other hormone-sensitive cancers, such as breast, prostate and endometrial cancers, have led investigators to study hormonal therapy in women with ovarian cancer. Estrogen and progesterone receptors are reported in approximately 50% of tumors and 98% of ovarian cancer contains androgen receptors[105-108].

Dose-dependent inhibitory effect of anti-androgens (anandron, flutamide and hydroxy flutamide) is observed in about 60% of primary cultures of ovarian cancer in vitro. Progesterone and tamoxifen inhibited cell proliferation in only about 15% [105,109].

1.3.7 Prognosis

The overall prognosis of ovarian carcinoma remains poor, a direct result of its rapid growth rate and the lack of early symptoms. The survival rate is approximately 35% at 5 years, 28% at 10 years and 15% at 25 years[2,110]. Factors known to influence prognosis are listed below.

1. *Age.* As a group, younger patients have a better outcome. This is at least partially due to the fact that there is a higher percentage of borderline, well-differentiated and stage I tumors in this group[111, 112].

2. *Ascites.* This clinical finding constitutes, by itself, an unfavorable prognostic sign [113].

3. ***Borderline versus invasive tumor.*** This is a distinction of utmost significance for prognostic purposes. The incidence of recurrence is nearly zero for borderline mucinous and endometrioid tumors and about 20% for borderline serous or seromucinous tumors, the survival was 99% for stage I tumor and an astonishing 92% for advanced stage disease [114-116].

4. ***Tumor grade and type.*** Among the carcinomas, tumor grad correlates closely with survival. As a group, endometrioid carcinomas do better when pure than when mixed with papillary serous or undifferentiated components [117-119].

5. ***Psammoma bodies.*** Serous tumors containing numerous such structures have a better prognosis. This may be related to the fact that most of these tumors are well differentiated [120].

6. ***Rapture of tumor capsule.*** There is no convincing evidence that this occasional intraoperative complication has any influence on survival rates[121].

7. ***DNA ploidy.*** DNA analysis with flow cytometry has proved a strong prognostic indicator, in the sence that aneuploid tumors belong to a higher grade and behave in a much more aggressive fashion than the

diploid tumors. Correlation has also been found between tumor DNA ploidy and response to chemotherapy [122-124].

8. *CA-125.* Serum CA-125 levels frequently reflect the volume of disease and as such, in multivariate analysis, preoperative levels have failed to exert an independent prognostic effect on survival [125].

1.3.8 Screening

Epithelial ovarian cancer contineu to challenge clinicians, there is no accepted method for screening this cancer and most patients present with advanced disease[61]. It has been suggested that to improve early diagnosis, screening efforts should be directed to specific groups of high risk young patient, such as those with family history of ovarian or breast cancer and nulliparous or infertile women undergoing induction of ovulation, or 1st degree relative who has one of the hereditary syndromes[33, 125]. The lack of appropriate specificity for serum CA-125 provides substantial evidence against its use as screening tool for ovarian cancer. Furthermore, normalization of serum CA-125 is not guarantee of freedom from cancer[126-127]. *Trans–vaginal ultrasound (TV U/S)* shown to be useful in evaluating ovarian mass for malignant potential. Combined use of CA-125 and trans-vaginal ultrasound may increase specificity over either method used independently, in addition these tests are not cost-effective and therefore are not recommended for use in routine screening for this disease[128-130].

1.4 Glycoprotein Hormones

The glycoprotein hormones of the pituitary (**LH, FSH, TSH**) and of the placenta (**hCG**) are composed of two peptide chains, usually referred to as α and β subunits, each with carbohydrate substituent group attached. The carbohydrate moiety, which accounts for 15 to 31% of the molecular weight includes L-fucose, D-glucose, D-mannose, N-acetyl glucose amine, N-acetyl galactose amine and sialic acid [130,131].

Even with homology of amino acid sequence, major differences in carbohydrate composition exist. Also, there is microheterogeneity of the carbohydrate components of a single hormone, which suggests that the addition of these components to the peptide chain is not under rigorous metabolic control [132]. The oligo saccharides attached to the glycoprotein hormones are not required for binding to hormone receptors but are essential for activation of adenylate cyclase, and for the determination of the half- life of the hormones in circulation [133,134].

The α subunit for each hormone is almost identical, but the β- subunit differ considerably from one hormone to another, this suggests that the β-subunits carry the hormonal specificity. The β- subunit is not active by itself, and receptor recognition involves the interaction of regions of both subunits, the linkes between the two subunits are non covalent[135-138].

The carboxyl terminal pentapeptide of α is essential for receptor binding but not for α/β association. The feature that distinguishes hormones in the glycoprotein group from hormones of other groups is their glycosylation, in each glycoprotein hormone, the α subunit contains two complex asparagine-linked oligosaccharides, and the β subunit has either one or two. The glycosylation may be necessary for α/β interaction. The α subunit has five s-s bridges, and the β moiety has six [139].

The α and β chains of glycoprotein hormones are produced as precursors, from which a signal peptide is cleaved immediately after secretion into the lumen of the endoplasmic reticulum, subsequent processing, including glycosylation and possibly proteolytic modification at the N-terminus, occurs while the hormone passes through the secretory pathway, particularly in the golgi body. The hormones are stored as secretory granules which are mostly 250-300 nm in diameter[140,141].

Glycoprotein hormones are water soluble molecules circulate in the watery blood plasma in a "free" form, not attached to plasma proteins and they are consist of chains of 200-230 amino acids. The half life of glycoprotein hormones is short, varying between 8 and 60 min. A percentage of intact hormone and fragments of hormone is filtered by the kidney and excreted in the urine, they are also degraded by proteases and peptidases in plasma and at the target gland after internalization other organs such as the liver and lungs, which also metabolize glycoprotein hormones[9,40,142,143]. Hormones with short half-lives normally have concentrations that increase and decrease rapidly within the blood, they generally regulate activities that have a rapid onset and a short duration[6].

1.5 Gonadotrophin LH

Luteinizing hormone (Luteotropin) is produced in the gonadotrophic cells of the pituitary. This hormone is responsible for the gametogenesis and steroidogensis in the gonads[139, 143].

The molecular weight of LH is about 28KD, the amino acid sequence of the α and β subunits of LH were shown in figure (1-4). Expected values for LH in serum is as shown in table (1-3)[145].

Table (1-3): Expected values in serum for LH in normal adults [145]

	LH (mIU/ml)
Males, 23-70 years	1.2-7.8
Females	
Follicular phase	1.7- 15.0
Midcycle peak	21.9-56.6
Luteal phase	0.6- 16.3
Postmenopausal >50 year	14.2-52.3

a. HOOC-Ala-Pro-Asp-Val-Gln-Asp-Cys-Pro-Glu-Cys-Thr-Leu-Gln-Glu-Asn-Pro-Phe-Phe-Ser-Gln-Pro-Gly-Ala-Pro-Ile-Leu-Gln-Cys-Met-Cly-Cys-Cys-Phe-Ser-Arg-Ala-Tyr-Pro-Thr-Pro-Leu-Arg-Ser-Lys-Lys-Thr-Met-Leu-Val-Glu-Lys-Asn-Val-Thr-Ser-Glu-Ser-Thr-Cys-Cys-Val-Ala-Lys   -Ser-Tyr-Asn-Arg-Val-Thr-Val-Met-Gly-Gly-Phe-Lys-Val-Glu-Asn-His-Thr-Ala-Cys-His-Cys-Ser-Thr-Cys-Tyr-His-Lys-Ser-OH

b. H-Ser-Arg-Glu-Pro-Leu-Arg-Pro-Trp-Cys-His-Pro-Ile-Asn-Ala-Ile-Leu-Ala-Val-Glu-Lys-Glu-Gly-Cys-Pro-Val-Cys-Ile-Thr-Val-Asn-Thr-Thr-Ile-Cys-Ala-Gly-Tyr-Cys-Pro-Thr-Met-Arg-Val-Leu-Gln-Ala-Val-Leu-Pro-Pro-Leu-Pro-Gln-Val-Cys-Thr-Tyr-Arg-Asp-Val-Arg-Phe-Glu-Ser-Ile-Arg-Leu-Pro-Gly-Cys-Pro-Arg-Gly-Val-Asp-Pro-Val-Val-Ser-Phe-Pro-Val-Ale-Leu-Ser-Cys-Arg-Cys-Gly-Pro-Cys-Arg-Arg-Ser-Thr-Ser-Asp-Gys-Gly-Pro-Lys-Asp-His-Pro-Leu-Thr-Cys-Asp-His-Pro-Glu-NH$_2$

Figure (1- 4): Amino acid sequences of the: [130, 135]
(a) α subunit of hLH
(b) β subunit of hLH

It is clear from the influence of sex and age on LH secretion that a patient serum value must be compared to normal values obtained from individuals of the same sex and age, and in females, preferably at the same phase of the menstrual cycle [23].

1.5.1 Mechanism of Action of Luteinizing Hormone (LH)

Protein hormones are too polar to diffuse passively through lipoprotein membranes and also too large to pass through membrane pores, instead these hormones initiate their response by binding to receptor located on or in the cell membrane. This binding interaction results in the generation of an intracellular signal or "second messenger" that in turn mediates the hormone's effect on intracellular enzymes, gene expression and membrane transport. Whereas the "first messenger" of intracellular communication is the hormone, the second messenger may be a small organic molecule, such as cyclic adenosine monophosphate (cAMP) or inositol phosphate, it may be an ion such as calcium; or it may be the activation of a protein kinase [9.146-148].

Luteinizing hormone plays a key roles in the control of reproductive function and act on specific receptors in the cell membrane of target cells in the gonads. LH receptor together with FSH, TSH receptors are members of the super family of G protein coupled receptors. These receptors use guanine nucleotide-binding proteins (G protein) to interface with target proteins, such as adenylate cyclase. Receptors of this class are monomeric proteins with an extracellular domain that binds the hormone, a cytoplasmic domain that binds the G protein, and seven trans membrane domains [148-151].

The action of a typical water- soluble hormone (Figure 1-5) occurs as follows:[9]

1. A water- soluble hormone diffuses from the blood through interstitial fluid and then binds to its receptor in a target cell's plasma membrane, this binding activates another membrane protein, called a G- protein which activates adenylate cyclase.

2. Adenylate cyclase then converts Adenosine Tri Phosphate (ATP) into cyclic AMP (cAMP) in the cytosol of the cell.

3. Cyclic AMP (the second messenger) activates one or several protein kinases, which may be free in the cytosol or bound to the plasma membrane. A protein kinase is an enzyme that phosphorylates (add a phosphate group to) cellular proteins. The donnar of the phosphate group is ATP, which is converted to ADP.

4. Activated protein kinases phosphorylate one or several other enzymes, here called enzyme 1, and enzyme 2, phosphorylation activates some enzymes and inactivates others, rather like an on-off switch. The result

of phosphorylating a particular enzyme could be regulation of other enzymes, secretion, protein synthesis or a change in plasma membrane permeability.

Enzyme activated by phosphorylation, in turn, catalyze reactions that produce physiological responses. After a brief period of time, cyclic AMP activity regulated by the cytoplasmic enzyme phosphodiesterase, which deactivates cyclic AMP by converting it to $\bar{5}$-adenosine monophosphate[40].

1.5.2 Hormonal Control

1.5.2.1 Hypothalamic-Pituitary-Ovarian Axis

Growth, prepubertal maturation, reproductive cycle and sex hormones secretion are regulated by *FSH* and *LH* from the anterior pituitary glands. The secretion of both LH and FSH is stimulated by *gonadotrophin releasing hormone (GnRH)* from the hypothalamus. This hormone has been purified and found to be a deca peptide with the following formula:[1,3,68,152]

Glu-His-Trp- Ser- Tyr- Gly- Leu- Arg- Pro- Gly-NH2

The release of GnRH by the hypothalamus is influenced by neurons from other regions of the brain whose terminals end in the arcuate nucleus[153, 154]. Because LH and FSH are secreted in pulsatile bursts, it follows that GnRH is also pulsatile. The pulsatile release of GnRH is required but plays only a permissive role in the mid-cycle surge of gonadotrophins which is regulated primarily by ovarian hormones feed back at the level of pituitary [155].

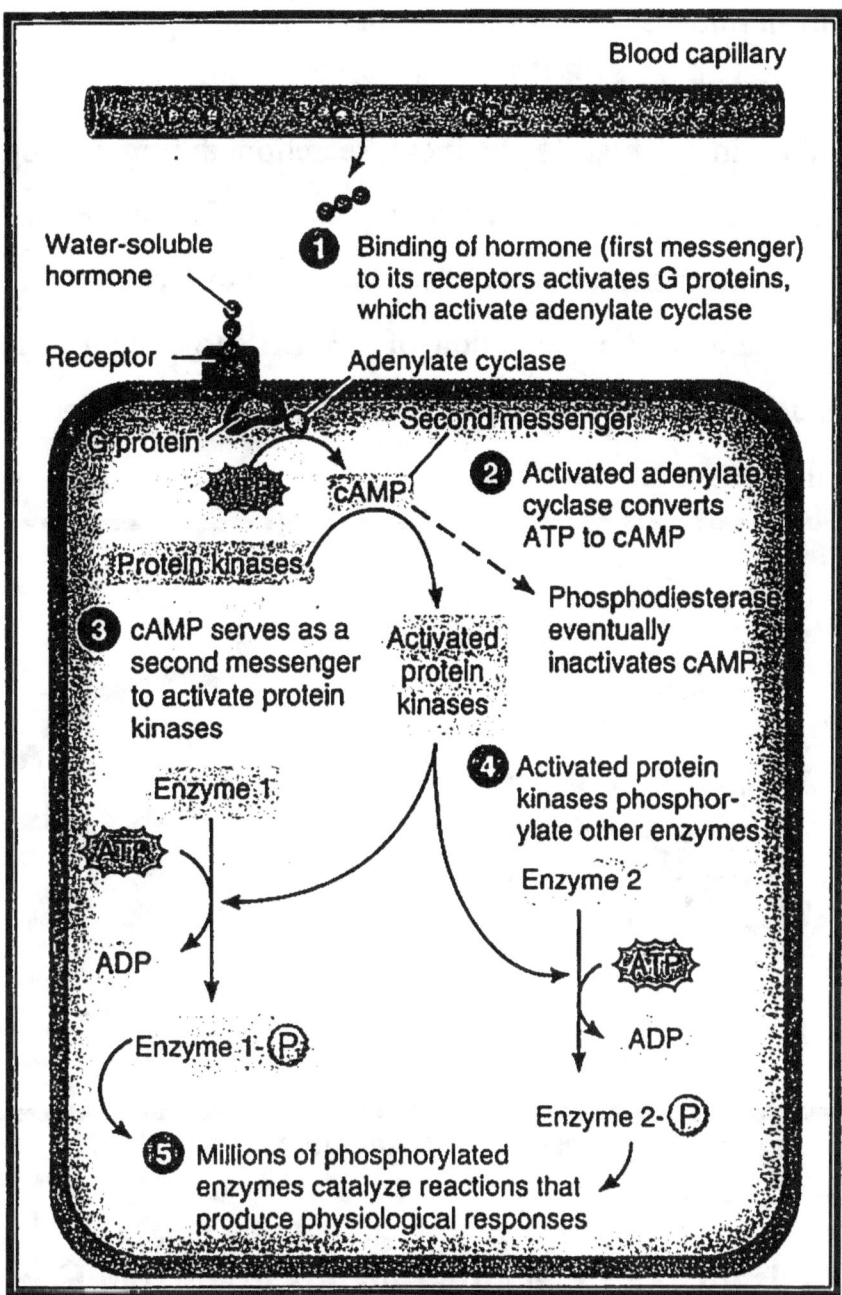

Figure (1-5): Mechanism of action of the water-soluble hormones[9].

1.5.2.2 Feed Back Mechanism

The release of LH is affected both *positively* and *negatively* by estrogen and progesterone (Figure 1-6). Whether estrogen and progesterone stimulate or inhibit LH release depend upon the level of exposure and duration of the steriod receptors for estrogens and progesterone are found in various parts of the brain, including the hypothalamus. Ovarian steriod and peptide hormones can exert a negative feed back on both the hypothalamus and pituitary[156,157].

A decline in ovarian hormones secretion during menopause or following castration causes increased secretion of LH and FSH. Ovarian steriod and peptide hormones are also able to exert positive feed back which is important in the regulation of the LH surge required to induce ovulation and is regulated by sharply rising levels of estrogen in the late first half of the menstraul cycle [6,8].

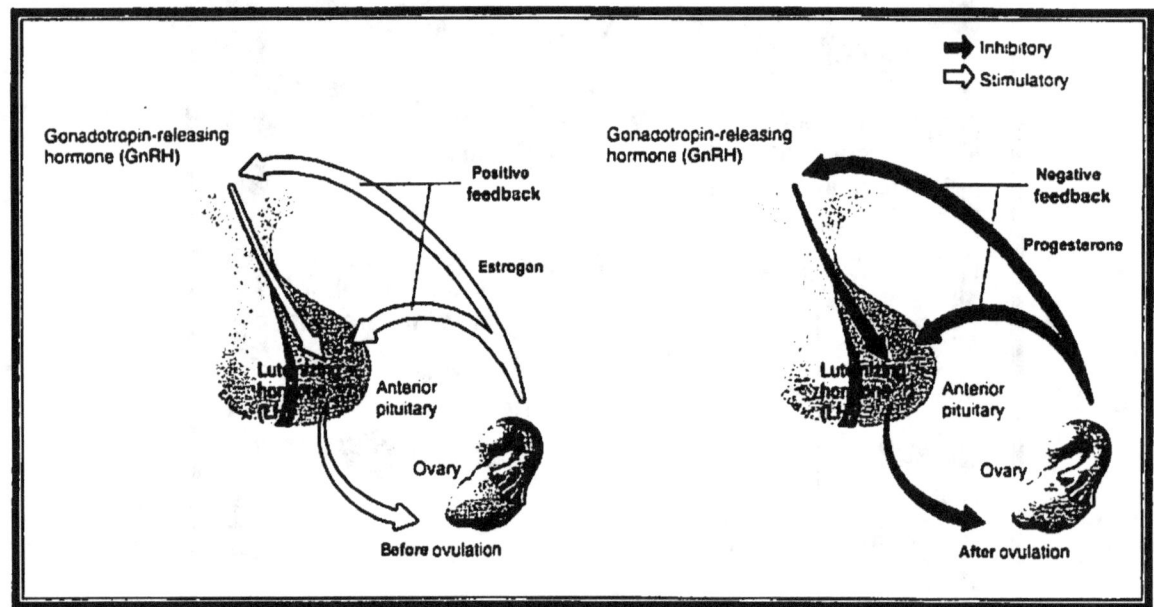

Figure (1-6): Positive and negative feedback[6].

1.5.3 Methods for the Determination of Luteinizing Hormone

Methods for assaying gonadotrophins have evolved from *bioassay* to *receptor assays* to *immunoassays* that make use of radioactive and non isotopic labeles. Early in vivo bioassay were based on the effect of gonadotrophin on the rat ovary such as depletion of ascorbic acid by LH or augmentation of ovarian weight by FSH. These methods were insensitive and lacked the necessary specificity and precision to measure LH in small volumes of serum; consequently, they were mainly applied to the assay of urine concentration [158,159].

Receptor assays are the most important examples of competitive binding assays. Binding of the hormone to a specific macromolecule (usually a protein) is used as the basis of the method, competition for binding between the hormone sample under study and the radioactivily labeled hormone being followed, then separate the receptor-bound and free hormone and determination of labeled hormone bound relating this to the amount of unlabeled hormone present using a standard curve [160, 161].

Radioimmunoassay is another type of immunoassays based on the reaction between a labeled hormone and an antibody ligand, it is assumed that a protein hormone reacts with only one species of antibody, at a single binding site. The reaction leading to binding may be designated as follows:

Hormone + Antibody ⇌ Hormone antibody complex

If labeled hormone is added to such a system, then both labeled and unlabeled hormone will compete for the same binding site, therefore, the greater concentration of unlabeled hormone, the less radioactivity will be bound to a limited quantity of antibody [159,162,163], and to separate free from bound hormone, double antibody precipitation techniques is usually used. In these techniques, the antibody is generated against the globulins of one species in another species, and the anti-gamma globulins are so made to precipitate immunoglobulins, including those which bind hormones, such a precipitate may be filtered off, or alternatively centrifuged down. The concentration of gonadotrophins in the sample is determined by comparing radioactivity in the specimen tubes with the calibration curve prepared with known concentrations of Gn[164-166].

An interesting variant of radioimmunoassay procedures employs the use of iodine-labeled antibody, such methods have been called *immunoradiometricassays (IRMA)*, the introduction of two site immunometricassays has been responsible for the marked improvements in gonadotrophin measurements. A number of LH "Sandwich" assays are available for manual or automated applications [148].

A simultaneous ***immunofluorometric assay*** of LH based on the use of the fluorescent lanthanides Eu^{3+} and Tb^{3+} each is detected in a time resolved fluorometric [167].

Enzymes may also be used in a similar way, these are coupled to a hormone by chemical methods. The enzyme-labeled material is then used in the assay as if it was isotopically labeled, ***an enzyme immunoassay (EIA)*** is frequently employed with a solid phase type of assay *(called **enzyme-linked immunoabsorbent assay or ELISA)***, the enzyme is detected by adding a suitable substrate and then measuring its conversion to some other chemically detectable material [138,163].

Aim of the Work

The aim of the work includes the following:

1. Determination of the Luteinizing hormone (LH) level in sera of normal postmenopausal women and patients affected by serous ovarian tumors.

2. Molecular characterization of the binding of ^{125}I-hLH with its receptors in malignant serous ovarian tumors of postmenopausal patients and benign serous tumors of both premenopausal and postmenopausal patients; such as those of binding capacity and the effect of various factors like (temperature, time, pH, Salts, halides, receptor concentration, hormone concentration).

3. Determination of the kinetic and thermodynamic parameters of the binding reaction of luteinizing hormone with its receptors in benign and malignant serous ovarian tumors.

4. Spectroscopic studies on the hLH, hLH antibody, hLH-antibody complex, ^{125}I-hLH binding with its receptors in ovarian tissue of patients with benign and malignant serous ovarian tumors.

Chapter 2

Experimental Work

2.1 Chemicals, Instruments and Samples

2.1.1. Chemicals

All laboratory chemicals and reagents were of analar grade and were used without further purification. These tris (hydroxy methyl)aminomethan, $CuSO_4.5H_2O$, NH_4Cl, $LiCl$, $CaCl_2$, $ZnCl_2$, $MnCl_2$, $MgCl_2$, dimethyl sulfoxide, Na, k-tartarate, were obtained from **Fluka company, Switzerland.**

NaCl, NaF, PEG-10,000, glycerol, hydrochloric acid, Na_2CO_3, NaOH, Folin ciocalteau, ethanol, and Bovine serum albumin (BSA) were obtained from **BDH limited pool, U.K.**

Blue dextran (2000) and sephadex G-100 (superfine) were obtained from **Pharmacia fine chemicals-Sweden.**

Kit of radioactive Luteinizing hormone (^{125}I-hLH) was purchased from **diagnostic products corporation (DPC)- United State.** The activity of labeled Luteinizing hormone was approximately 3μci (micro-curies).

2.1.2 Instruments

The instruments used in this work were, LKB gamma counter type 1270-rack gamma II, LKB spectrophotometer ultraspec type 4050, Pye-unicom pH meter, Varian DMS 100 UV- visible spectrophotometer, LKB ultracentrifuge type 2332, Memmert water bath, Memmert incubator.

2.1.3 Patients

One group of ovarian cancer patients and two groups of patients with benign ovarian tumors were included in this study. Group 1 contained 4 postmenopausal patients with epithelial ovarian cancer (serous cystadeno-carcinoma), group II consisted of 8 postmenopausal patients with benign epithelial ovarian tumor (serous cystadenoma) and group III consisted of 4 premenopausal patients with benign epithelial ovarian tumor (serous cystadenoma).

All patients were admitted for treatment to (**University Hospital, Saddam College of Medicine**), (**Saddam Medical City, Baghdad Teaching Hospital**), (**Al-Jamiaa Private Hospital and Al-Jaderia Private Hospital**) under the supervision of specialists **Dr. Tahrer Wade'e and Dr. Rajaa Al-Tekrety, Dr. Nada Al-Abady, Dr. Samya Al-Hashemy and Dr. Kammal Hussain.**

They were histologically proven from the supervision of specialists **Dr. Nawal Alash, Dr. Sahera Alwan, Dr. Awatof Al-Quraishee and Dr. Ala'a Abdul Gany;** newly diagnosed and not underwent any type of therapy. Patients suffered from any disease that may interfere with our study were excluded.

2.1.4 Blood Sampling

Blood samples (5-7 ml) were obtained from postmenopausal patients undergoing total abdominal hysterectomy with bilateral salpingo oöphorectomy by vein puncture before surgery. Age matched sera were obtained from (14) healthy postmenopausal women.

Hemolyzed, lipemic or icteric specimens should not be used. Blood samples were centrifuged at 1500 xg for 10 minutes after allowing the blood to clot at room temperature. Serum is preferred as the specimen of choice for gonadotropin. Luteinizing hormone is stable for 8 days at room temperature and for 2 weeks at 4°C; for longer periods, the serum specimen should be stored frozen at or below -20°C.[148]

2.1.5 Collection of Ovarian Tissue Specimens

The tumors tissues were surgically removed from ovarian tumor patients by either unilateral salpinco oöphorectomy or total abdominal hysterectomy with bilateral salpingo oöphorectomy. The specimens were cut off and immediately rinsed with ice-cold isotonic saline solution. They were collected individually in plastic receptacles and stored at -20°C until homogenization.

2.1.6 Preparation of Ovarian Tumor Tissue Homogenate

The frozen tissues were weighed, pulverized finely with a scalpel in petri dish standing on ice bath, and then homogenized at 4°C in buffer solution with a ratio of 1:5 (weight:volume), using a manual homogenizer. The buffer used was sucrose-Tris (ST) buffer (0.01 M, pH 7.4).

The homogenate was filtered through several layers of nylon gauze to eliminate fibers of connective tissue, then centrifuged at 1000 xg for 15 minutes at 4°C in order to precipitate the remaining intact cells and nuclau. The supernatant was separated, divided in aliquots and frozen until the time of the experiments.

2.1.7 Solutions

ST buffer: sucrose-Tris buffer (0.01 M, pH 7.4) was prepared by dissolving (0.606 gm) of Tris (hydroxy methyl) aminomethane and (51.345 g) of sucrose in (450 ml) distilled water (D.W.). Then the pH was adjusted to pH 7.4 using HCl (0.2 M). The volume was made up to (500 ml) with D.W.

2.2 Binding Studies of ^{125}I-Luteinizing Hormone With Its Receptors in Serous Ovarian Tumor Homogenate

2.2.1 Determination of Protein in Serous Ovarian Tumor Homogenate

The total protein content of serous ovarian tissue homogenate was determined by the method of *Lowry et. al.*, using bovine serum albumin (BSA) as the standard protein[168].

Procedure

1- Zero, 10, 20, 40, 60, 80, 100, 120, 150, 200 μl of standard BSA (0.2 mg/ml) were pipetted in a set of test tubes. Volumes were made up to 1 ml with

D.W. to give final concentration of zero, 10, 20, 40, 60, 80, 100, 120, 150 and 200 μg/ml protein. The experiment was carried out in duplicate.

2- A volume of (25 μls.) of serous ovarian tissue homogenate (benign and malignant tumors) was pipetted in a test tube and the volume was made to 1 ml with D.W.

3- Five milliliter of reagent C was added to all assay tubes. Then the contents were shaked and allowed to stand for 10 minutes at room temperature.

4- A volume of 0.5 ml reagent D was added to all assay tubes and mixed immediately. The mixture was left to stand for 30 minutes at room temperature.

5- The absorbance of the developing color was read at 750 nm against the appropriate blank.

6- The standard curve was obtained by plotting the absorbance against the corresponding concentrations of standard protein and used to determine the unknown protein concentration of the homogenate of ovarian tumors, figure(2-1).

Reagents

1- Standard bovine serum albumin (BSA) 0.2 mg/ml as stock solution.

2- Reagent A, Alkaline sodium carbonate solution:(2% Na_2CO_3 in 0.1 N NaOH).

3- Reagent B, Copper sulphate-sodium potassium tatrate solution: (0.5% $CuSO_4.5H_2O$ in 1% Na, k tartrate).

4- Reagent C, Alkaline copper solution (50 ml of reagent A and 1 ml of reagent B), discard after one day.

5- Reagent D, Folin ciocalteau reagent (1N): prepared by the dilution of the commercial reagent (2N) with an equal volume of D.W. on the day of use.

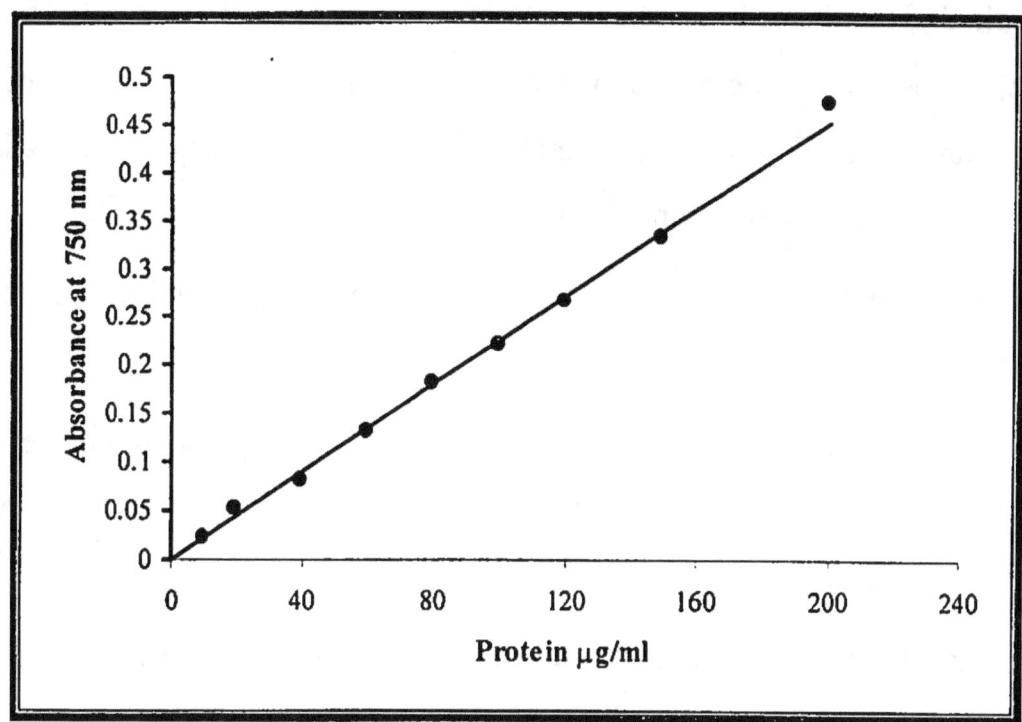

Figure (2-1): Standard curve for protein determination by *Lawry's* method [168].

2.2.2 Determination of ^{125}I-Luteinizing Hormone Concentration

The concentration of labeled Luteinizing hormone was measured according to the method of *Morris*[169].

The method is outlined according to the following steps:

1- Two hundred microliters of each standard of unlabeled hormone of different concentration (0, 3, 10, 20, 40, 100 and 200 mIU/ml) were pipetted in a set of tubes marked from 1 to 14 and according to the assay protocol described in table (2-1).

2- Anti-LH serum (100 µls.) was added to all the assay tubes. The tubes were vortex-mixed for few minutes until the color should be light blue.

3- All the assay tubes were stoppered and incubated for 30 minutes at 37°C.

4-   One hundred microliters of ^{125}I-hLH was pipetted into the assay tubes marked from 1 to 14 and volumes of ^{125}I-hLH ranging from 150-350µls. were pipetted in another set of tubes marked from 15 to 24, 200µls. of the standard containing zero concentration of unlabeled hLH, was also added to all the tubes marked from 15 to 24.

5-   All the assay tubes were stoppered and incubated for 60 minutes at room temperature.

6-   To prepare total count, the same amounts of ^{125}I-hLH prepared in step (4) were pipeted into a separate set of tubes and the radioactivity was determined using gamma counter.

7-   One milliliter of the second antibody reagent was pipetted into all the reaction tubes. The tubes were then vortex-mixed and allowed to stand for at least 5 minutes.

8-   All the assay tubes were centrifuged for 30 minutes at 1500 xg in a cooling centrifuge (4°C).

9-   After centrifugation, the tubes were placed carefully in a suitable decanting racks where the tubes were kept inverted and placed on a pad of cotton after the discard of the supernatant. This step lasts at least for 10 minutes in order to allow the complete drainage of any last drops of supernatant liquid.

10-   The rims of the tubes were blotted with cotton sticks to remove any persistent droplets of the supernatant. Care should be taken not to remove the precipitate in the bottom of each tube.

11-   All the tubes were stoppered and the radioactivity was counted for 1 minutes using gamma counter.

Table (2-1): Assay protocol for determination ^{125}I-hLH concentration (mIU/ml) that use in binding experiments.

	Unlabeled LH standard concentration (mIU/ml)											
	0	3	10	20	40	100	200	0	0	0	0	0
Tube number	1,2	3,4	5,6	7,8	9,10	11,12	13,14	15,16	17,18	19,20	21,22	23,24
Standard concentration	200	200	200	200	200	200	200	200	200	200	200	200
anti LH serum	100	100	100	100	100	100	100	100	100	100	100	100
vortex mix, incubated 30 minutes at 37°C												
^{125}I-hLH	100	100	100	100	100	100	100	150	200	250	300	350
vortex mix, incubated 60 minutes at room temperature												
second antibody reagent	1000	1000	1000	1000	1000	1000	1000	1000	1000	1000	1000	1000
vortex mix, leave at room temperature for at least 5 minutes and centrifuge for 30 minutes at 1500 xg.. Decant supernatant solutions. Measure radioactivity in precipitates.												

All volumes are in microliters.

Reagents

The reagents provided in the LH-RIA kit from DPC, USA were used:

1- Radio labeled LH (^{125}I-hLH): One vial of lyophilized iodinated LH, reconstituted by adding 10 ml distilled water.

2- LH calibrators: Seven vials of hLH calibrators supplied in liquid form ready to use, each vial contains 3.0 ml of different concentrations of unlabeled hormone (3, 10, 20, 40, 100 and 200 mIU/ml) and one vial contains 6 ml of zero concentration.

3- Anti LH serum: One vial of lyophilized LH anti serum, reconstituted by adding 10 ml distilled water.

4- Precipitating solution: One vial containing (110 ml) of precipitating solution, consisting of goat anti rabbit gamma globulin and dilute polyethylene glycol in saline.

Calculations

1- (B) is the bound radioactivity (CPM) which represents the counted radio-activity in the precipitated hormone-antibody complex.

2- The free hormone (F) which represents unbound ^{125}I-hLH was determined from the following formula:

   F(CPM)= Total count (CPM) - Bound radioactivity (CPM)

3- The values of the ratio (B/F) for an ordinary standard curve (displacement curve) were calculated. This standard curve represents the incubation of different amounts of hLH standards with constant amount of ^{125}I-hLH (Table 2-2).

4- The B/F values for the incubation of different amounts of ^{125}I-hLH in the presence of zero standard only were also calculated (Table 2-3).

5- The data in tables (2-2) and (2-3) were plotted as in figure (2-2).

6- Using the two curves I and II in figure (2-2), we can get the amount of the radioactivity corresponding to the concentration of unlabeled hormone (Table 2-4). This was done by drawing a line which intersects with both two curves at the same increment as shown in figure (2-2).

7- The plot of the data in table (2-4) results in a straight line, the increment on the ordinate represents the concentration of the tracer ^{125}I-hLH in mIU/ml (Figure 2-3).

Table (2-2): B/F values corresponding to different concentration of standard hLH used in standard curve.

Concentration of standard hLH (mIU/ml)	B/F
0	0.45
3	0.40
10	0.331
20	0.245
40	0.154
100	0.082
200	0.0436

Table (2-3): B/F values corresponding to different amounts of ^{125}I-hLH used in the incubation.

Amount of ^{125}I-hLH (CPM)	Bound radioactivity (CPM)	B/F
12550	3650	0.41
16662	4500	0.37
20554	5100	0.33
26286	5750	0.28
31250	6250	0.25

Table (2-4): The mass of the standard hLH in (mIU/ml) corresponding to a given amount of the tracer.

Bound radioactivity (CPM) × 10³	Amount (mIU/ml)
4.0	3
4.5	6
5.0	9
5.5	13
6.0	17

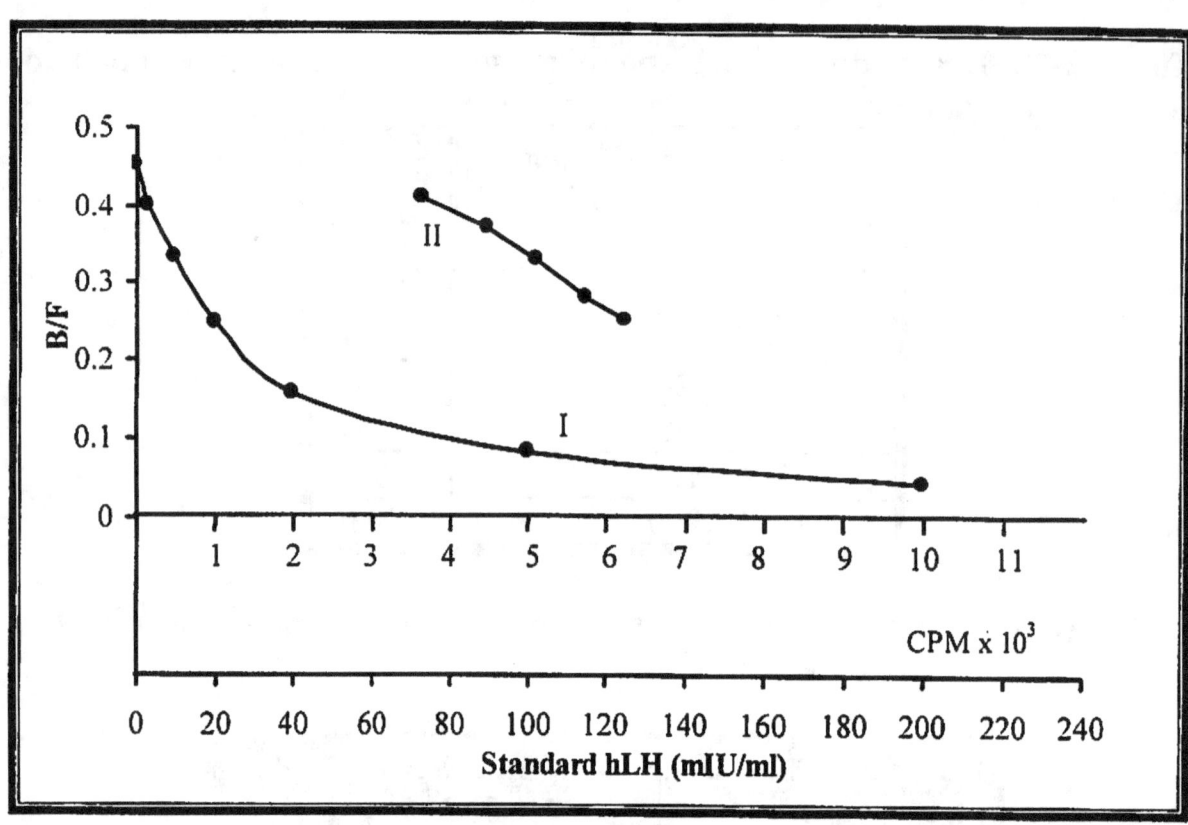

Figure (2-2): Ratio of bound to free radioactivity (B/F) for an ordinary standard curve, where different amounts of LH standard were incubated with constant amount of ^{125}I-hLH and antibody (I), also B/F for antibody incubated with different amounts of ^{125}I-hLH in the absence of unlabeled LH (II).

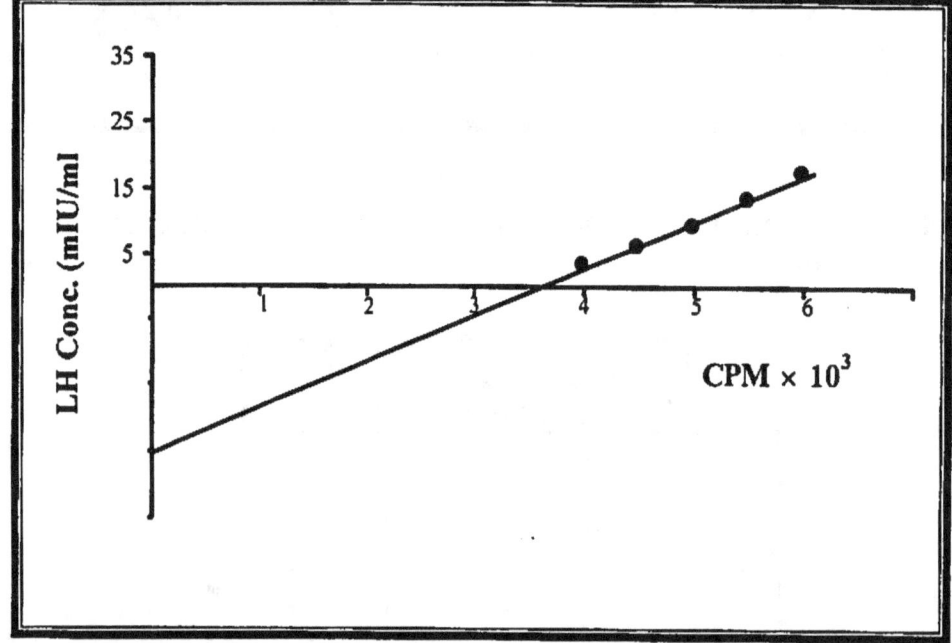

Figure (2-3): A plot of the mass of standard LH against the CPM of ^{125}I-hLH having the same B/F from fig (2-2), which resulted a straight line. The intercept on the ordinate corresponds to the concentration of ^{125}I-hLH in mIU/ml.

2.2.3 Determination of Luteinizing Hormone Levels in Sera of Serous Ovarian Tumor Patients and Controls

Serum levels of LH were measured on samples collected from postmenopausal patients and healthy postmenopausal women by radioimmuno-assay (RIA).

The assay protocol was described in table (2-5) according to the following steps:

1- Label sixteen tubes in duplicate: NSB (non specific binding), A (maximum binding MB) and B through G for different concentration of standard LH. Label additional tubes, also in duplicate, for serum samples and controls.

2- Two hundred μls. of the zero standard was pipetted into NSB and A(MB) tubes, and 200 μls. of each of the remaining standards B through G into correspondingly labeled tubes. Two hundred μls. of each patient serum sample and control were pipetted into the tubes prepared.

3- One hundred μls. of anti LH serum was added to all tubes except the NSB tubes.

4- All the assay tubes were stoppered and incubated for 30 minutes at 37°C.

5- One hundred μls. of ^{125}I-hLH was pipetted to all the assay tubes, then all the assay tubes were stoppered and incubated for 60 minutes at room temperature.

6- The radioactivity of (^{125}I-hLH-anti LH serum) complex formed was estimated by following the steps 7, 8, 9, 10 and 11 in section (2.2.2).

Table (2-5): RIA assay protocol of serum LH(mIU/ml)

	NSB	A(MB)	B	C	D	E	F	G	Control		unknown		
unlabeled LH standard concentration (mIU/ml)	0	0	3	10	20	40	100	200	-		1	2	3.. etc
tube number	1,2	3,4	5,6	7,8	9,10	11,12	13,14	15,16	17,18	19,20	21,22	23,24	25,26
standards concentration	200	200	200	200	200	200	200	200	-	-	-	-	-
control serum or samples	-	-	-	-	-	-	-	-	200	200	200	200	200
anti LH serum	-	100	100	100	100	100	100	100	100	100	100	100	100
vortex mix, incubated for 30 minutes at 37°C													
^{125}I-hLH	100	100	100	100	100	100	100	100	100	100	100	100	100
vortex mix, incubated for 60 minutes at room temperature													
second antibody reagent	1000	1000	1000	1000	1000	1000	1000	1000	1000	1000	1000	1000	1000
vortex mix, leave at room temperature for at least 5 minutes and centrifuge for 30 minutes at 1500 xg. Decant supernatant solutions. Measure radioactivity in precipitates.													

All volumes are in microliters.

Reagents

The reagents used in this experiment were described in section (2.2.2).

Calculations

1- The mean net count for each group of tubes was counted in a gamma counter for 1 minute, then the corrected count per minute was calculated as follows:

Net count = Average (CPM)- Average NSB(CPM).

2- Bound radioactivity was determined for each standards and serum sample or control as a percent of maximum binding MB as follows:

$$\text{Percent Bound} = \frac{\text{Standard or sample net count}}{\text{Net MB counts}} \times 100$$

3- A standard curve was drawn by plotting the percent bound value for each standard against the corresponding LH standard concentration mIU/ml (in log-log coordinates), figure (2-4).

4- The concentration of the unknowns were calculated from the standard curve using the mean of their duplicate net counts.

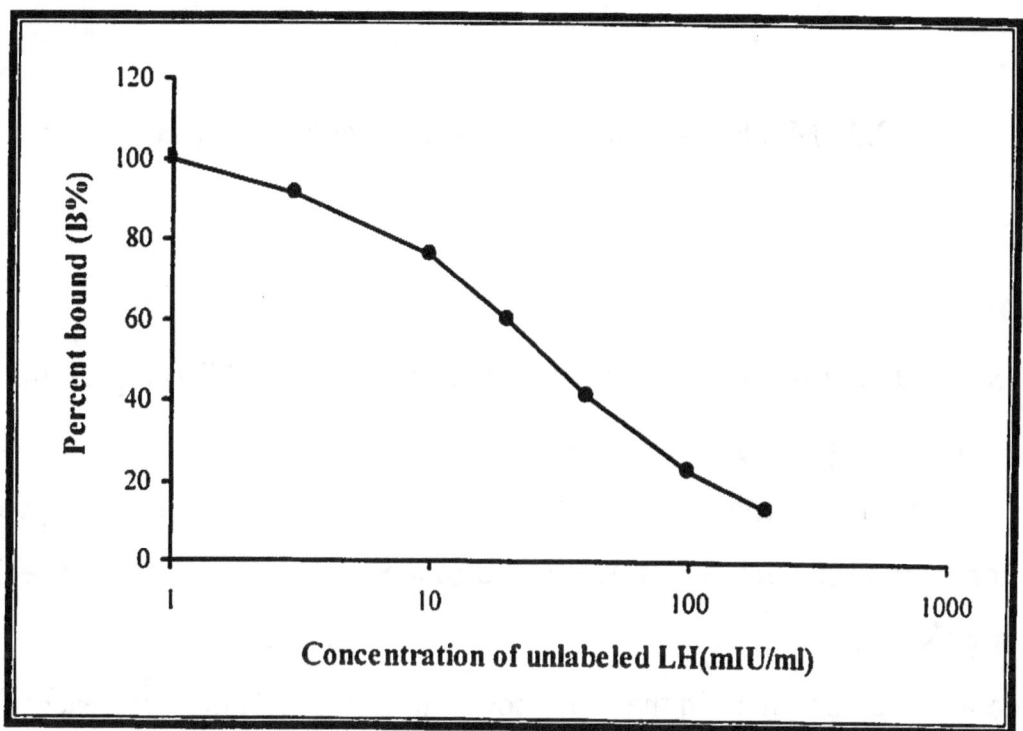

Figure (2-4): Typical plot of radioimmunoassay (RIA) for determining LH-levels in sera.

2.2.4 Preliminary Test of ^{125}I-hLH Binding to Its Receptors in Serous Ovarian Tumor Homogenate

1- A volume of 100 µls. of tissue homogenate were added to 50 µls. of labeled hLH (25.625 mIU/ml). The volume of the mixture was completed to 0.5 ml with ST buffer (0.01M, pH 7.4), then the tubes were stoppered and incubated for 120 minutes at 37°C. Non-specific binding was accounted by preparing the same incubation mixture with the addition of 20 fold excess of unlabeled hLH as a competitor. Two additional tubes containing 50 µls. of ^{125}I-hLH only (for total concentration of hormone CPM computation), were set a side until counting.

2- After incubation, the tubes were centrifuged at 1500 xg and 4°C for 45 minutes in order to separate the (^{125}I-hLH-receptor) complex.

3- The radioactivity of (^{125}I-hLH-receptor) complex formed was estimated by following the steps 9, 10 and 11 in section (2.2.2).

Solutions

ST buffer (0.01 M, pH 7.4) was prepared as described previously in section (2.1.7).

Calculations

1- The counted radioactivity in each tube (expressed in CPM) represents the total binding (TB).

2- The counted radioactivity (expressed in CPM) in the tubes contained labeled hormone and excess of unlabeled hormone represents the non-specific binding (NSB).

3- The counted radioactivity in the tubes containing ^{125}I-hLH only represents the total concentration of hormone CPM.

4- The specific binding (CPM) was calculated by subtracting the radioactivity (CPM) obtained in the presence of unlabeled hormone from that produced in the absence of unlabeled hormone.

$$SB(CPM) = TB\ (CPM) - NSB\ (CPM)$$

5- The percent of specific binding (SB%) can be calculated from the following formula:

$$SB\% = \frac{SB}{TC} \times 100$$

where:

SB: Specific binding (CPM)

TC: Total concentration of hormone (CPM)

2.2.5 Radio Receptor Assay Studies of ^{125}I-hLH Binding to Its Receptors in Benign and Malignant Serous Ovarian Tumors

2.2.5.1 The Effect of Human Luteinizing Hormone Receptor Concentration on the Binding in Serous Ovarian Tumor Homogenate

Note: The experiment was carried out in duplicate

1- Fifty microliter (25.625 mIU/ml) of ^{125}I-hLH was incubated with 100 μls. of increasing amounts (25, 50, 100, 150, 200 and 250 μg protein) of ovarian tumor homogenate in a final volume of 0.5 ml (completed with ST buffer 0.01M, pH 7.4) with or without the addition of 20 fold excess of unlabeled hLH. All the assay tubes were shacked, stoppered and incubated for 120 minutes at 37°C.

2- Two additional tubes, containing 50 μls. (25.625 mIU/ml) of the ^{125}I-hLH only, for total concentration of hormone (CPM) computation, were set a side until counting.

3- At the end of incubation, all the assay tubes were centrifuged for 45 minutes at 1500 xg in a cooling centrifuge (4°C).

4- The radioactivity of (^{125}I-hLH-receptor) complex formed was estimated by following the steps 9, 10 and 11 in section (2.2.2).

Solutions

ST buffer (0.01 M, pH 7.4) was prepared as described previously in section (2.1.7).

Calculations

1- The percent of specific binding (SB%) was determined according to section (2.2.4).

2- The percent of specific binding (SB%) was plotted against the amount of protein receptors included each mixture.

2.2.5.2 The Effect of Different Concentration of ^{125}I-hLH on the Binding With Its Receptors in Serous Ovarian Tumor Homogenate

Note: The experiment was carried out in duplicate

1- One hundred μls. (150 μg protein) of benign and malignant postmenopausal ovarian tumors homogenates and one hundred μls.(200μg protein) of benign premenopausal ovarian tumor homogenate were added to increasing concentration of ^{125}I-hLH (0.1666- 3.332 nM; 5-100 μls.) with a final volume of 0.5 ml (completed with ST buffer 0.01M, pH 7.4), with or without the addition of 20 fold excess of unlabeled hLH. All the assay tubes were shacked, stoppered and incubated for 120 minute at 37°C.

2- Two additional tubes, containing increasing concentrations (0.1666- 3.332nM) of ^{125}I-hLH for total concentration of hormone (CPM) computation, were set aside until counting.

3- At the end of incubation the assay tubes were centrifuged for 45 minutes at 1500 xg in a cooling centrifuge (4°C).

4- The radioactivity of (^{125}I-hLH-receptor) complex formed was estimated by following the steps 9, 10 and 11 in section (2.2.2).

Solutions

ST buffer (0.01 M, pH 7.4) was prepared as described previously in section (2.1.7).

Calculations

1- the percent of specific binding (SB%) was determined according to section (2.2.4).

2- The percent of specific binding (SB%) was plotted against the different concentration of ^{125}I-hLH.

2.2.5.3 The Effect of Different pH on the Binding of ^{125}I-hLH to Its Protein Receptors in Serous Ovarian Tumor Homogenate

Note: The experiment was carried out in duplicate.

1- One hundred microliters (150 and 200μg) of the homogenate proteins were added to 25 μls. (0.833 nM) of ^{125}I-hLH in Benign and malignant postmenopausal ovarian tumors homogenates and to 50 μls. (1.666 nM) of ^{125}I-hLH in Benign premenopausal ovarian tumor homogenate with or without the addition of 20 fold excess of unlabeled hLH. The volumes of the mixtures were made up to o.5 ml with citric acid/phosphat buffer (0.01 M) of different pH ranging from (5.4-6.6) and with ST buffer (0.01 M) of different pH ranging from (7.0-9.0).

2- Two additional tubes containing 25 μls. (0.833 nM) and 50 μls. (1.666 nM) of the labeled hormone only for total concentration of the hormone (CPM) computation, were set a side until counting.

3- All the assay tubes were stoppered and incubated for 120 minutes at 37°C. At the end of incubation period, all the assay tubes were centrifuged for 45 minutes at 1500 xg in a cooling centrifuge (4°C).

4- The radioactivity of (^{125}I-hLH-receptor) complex formed was estimated by following the steps 9, 10 and 11 in section (2.2.2).

Solutions

1- ST buffer (0.01 M) of different pH was prepared as shown in section (2.1.7).

2- Citric acid/ phosphate buffer was prepared as follows:

Solution A: 0.01 M Citric acid (0.21 g $C_6H_8O_7$. H_2O) in 100 ml distilled water.

Solution B: 0.01 M Disodium phosphate (0.178g $Na_2HPO_4.12H_2O$) in 100 ml distilled water.

Working buffer pH (5.4-6.6) were prepared by mixing appropriate volumes of solution A and B to reach the required pH in a final volume of 100 ml.

Calculations

1- The percent of specific binding (SB%) was determined according to section (2.2.4) at each pH.

2- the percent of specific binding (SB%) was plotted against their corresponding pH values.

2.2.5.4 The Effect of Temperature on The Binding of ^{125}I-hLH to Its Receptors in Serous Ovarian Tumor Homogenate

Note: The experiment was carried out in duplicate

1- One hundred microliters (150 and 200 µg) of the protein in the homogenate were added to (25 and 50 µls.) of ^{125}I-hLH, with or without the addition of

20 fold excess of unlabeled hLH. The volume of the mixture were completed to 0.5 ml ST buffer pH 8.2 (postmenopausal benign and malignant tumors), and ST buffer pH 7.4 (premenopausal benign tumor). Two additional tubes containing (25 and 50μls.) of ^{125}I-hLH only for total concentration of hormone (CPM) computation, were set a side until counting.

2- All the assay tubes were stoppered and incubated for 120 minutes at 37°C.

3- At the end of incubation period, all the assay tubes were centrifuged for 45 minutes at 1500 xg in a cooling centrifuge (4°C).

4- The radioactivity of (^{125}I-hLH-receptor) complex formed was estimated by following the steps 9, 10 and 11 in section (2.2.2).

5- The experiment was repeated at different temperatures (4, 10, 25 and 45 °C).

Solutions

ST buffer (0.01 M, pH 8.2 and 7.4) were prepared as described in section (2.1.7).

Calculations

1- The percent of specific binding (SB%) was determined according to section (2.2.4) at each temperature.

2- The percent of specific binding values were plotted against the different temperatures of incubation.

2.2.5.5 The Choice of The Most Appropriate Incubation Time for The Binding of ^{125}I-hLH to Its Receptors in Serous Ovarian Tumor Homogenate

Note: The experiment was carried out in duplicate.

1- One hundred microliters (150 and 200 μg) of the protein in the homogenate were added to (25 and 50 μls.) of ^{125}I-hLH, with or without the addition of

20 fold excess of unlabeled hLH. The volume of mixture was completed to 0.5 ml with ST buffer (0.01M, pH 8.2 and 7.4). Two additional tubes containing (25 and 50 μls.) of labeled hLH only, for total concentration of hormone (CPM) computation, were set a side until counting.

2- The tubes were stoppered and incubated at 25°C (postmenopausal benign and malignant tumors homogenates) and 37°C (premenopausal benign tumor homogenate) at different time intervals (60, 90, 120, 150, 180, 240, 300 and 360 minutes).

3- At the end of incubation period, the assay tubes were centrifuged for 45 minutes at 1500 xg in a cooling centrifuge (4°C).

4- The radioactivity of (^{125}I-hLH-receptor) complex was estimated by following the steps 9, 10 and 11 in section (2.2.2).

Solutions

ST buffer (0.01M, pH 8.2 and 7.4) were prepared as described in section (2.1.7).

Calculations

1- The percent of specific binding (SB%) was determined according to section (2.2.4) at each time.

2- The percent of (SB%) was plotted against the different times of incubation.

2.2.5.6 Stability of (^{125}I-hLH-Receptor) Complex From Postmenopausal Serous Ovarian Tumors Homogenate

Note: The experiment was carried out in duplicate

1- The experiment was carried out at the optimum conditions of ^{125}I-hLH concentration, time, temperature and pH, in order to investigate the effect of temperature on (^{125}I-hLH-receptor) complex properties. The experiment was performed by adding (25 and 50 μls.) of ^{125}I-hLH to (150 and 200 μg) of the homogenate proteins with or without the addition of 20 fold excess of unlabeled hLH in a final volume of 0.5 ml (completed with ST buffer pH 8.2 and 7.4). The tubes were incubated at (25and 37°C) for (180 and 120 minutes) for (benign and malignant) postmenopausal and benign premenopausal ovarian tumors homogenates respectively.

2- After incubation, the bound hormone (hormone-receptor complex) was evaluated by following the steps 9, 10 and 11 in section (2.2.2).

3- The (hormone-receptor) complex was reincubated at different temperatures (0,25, 37 and 45°C). Between 0 and 8 hrs. the remaining bound hormone in each tube was measured by gamma counter.

Solutions

ST buffer (0.01 M, pH 8.2 and 7.4) were prepared as described previously in section (2.1.7).

Calculations

1- The (SB %) was estimated as mentioned in section (2.2.4).

2- The percent of specific binding (SB%) was plotted against the time of incubation.

2.2.5.7 Competitive Effect of Different Concentrations of Different Unlabeled Hormones on The Binding of ^{125}I-hLH to Its Receptors in Postmenopausal Serous Ovarian Tumor Homogenate

Note: The experiment was carried out in duplicate

1- The experiment was carried out at the optimum conditions, in order to establish binding site specificity, by determining the ability of different hormones (FSH, TSH and HCG) to compete with the radioligand for the binding sites. The experiment was performed by adding 25 μls. (25.625 mIU/ml) of ^{125}I-hLH to 100 μls. (150 μg protein) of benign and malignant ovarian tumors homogenates with or without the addition of 25 μls. of increasing concentration (20, 60 and 90 mIU/ml) of unlabeled hLH in a final volume of 0.5 ml (completed with ST buffer 0.01 M, pH 8.2).

2- Two additional tubes containing 25 μls. (25.625 mIU/ml) of ^{125}I-hLH only, for total concentration of hormone (CPM) computation, were set a side until counting.

3- The assay tubes were stoppered and incubated at 25°C for 180 minutes. At the end of incubation period, the radioactivity of (^{125}I-hLH-receptor) complex formed was estimated by following the steps 9,10 and 11 in section (2.2.2). Using a sample without the addition of any unlabeled hormone as a control.

4- The experiment was repeated with increasing concentrations (20, 60 and 90 mIU/ml) of unlabeled (FSH , TSH and HCG).

Solutions

ST buffer (0.01 M, pH 8.2) was prepared as described previously in section (2.1.7).

Calculations

1- The percent of specific binding (SB%) was estimated as mentioned in section (2.2.4).

2- The (SB%) was plotted against the concentrations of competitor (LH, FSH, TSH, HCG).

2.2.5.8 Effect of Different Halides on The Binding of ^{125}I-hLH To Its Receptors in Postmenopausal Serous Ovarian Tumor Homogenate

Note: The experiment was carried out in duplicate.

1- Twenty five microliters of ^{125}I-hLH (25.625 mIU/ml) was added to 100 μls. (150μg protein) of postmenopausal benign and malignant ovarian tumors homogenates with or without the addition of 20 fold excess of unlabeled hLH in a final volume of 0.5ml (completed with ST buffer 0.01 M, pH 8.2 containing 0.1 M of each of the following halides: NaF, NaCl, NaBr and NaI). A sample without the addition of any halides was used as a control.

2- Two additional tubes containing 25 μls. (25.625 mIU/ml)of ^{125}I-hLH only, for total concentration of hormone (CPM) computation, were set aside until counting.

3- The assay tubes were stoppered and incubated for 180 minutes at 25°C, then the radioactivity of bound hormone was estimated by following the steps 9, 10 and 11 in section (2.2.2).

Solutions

1- The halides stock solutions (0.1M) were prepared by dissolving each of the following amounts of salts in 250 ml ST buffer (0.01M, pH 8.2): 1.05 g of NaF, 1.461 g of NaCl, 2.572 g of NaBr and 3.75 g of NaI.

2- ST buffer (0.01M, pH 8.2) was prepared as mentioned in section (2.1.7).

Calculations

1- The (SB%) was estimated as mentioned in section (2.2.4).

2- The percent of specific binding was plotted against halides concentrations.

2.2.5.9 Effect of Mono and Divalent Cations on The Binding of ^{125}I-hLH With Its Receptors in Postmenopausal Serous Ovarian Tumor Homogenate

Note: The experiment was carried out in duplicate

1- Twenty five microliters of ^{125}I-hLH (25.625 mIU/ml) was added to 100 μls. (150 μg protein) of benign and malignant postmenopausal ovarian tumors homogenates with or without the addition of 20 fold excess of unlabeled hLH in a final volume of 0.5 ml (completed with ST buffer 0.01 M, pH 8.2 containing 25 mM of each of the following salts: (KCl, $LiCl$, NH_4Cl, $MgCl_2$, $MnCl_2$, $CuSO_4.5H_2O$, $ZnCl_2$ and $CaCl_2$). A sample without the addition of any salts was used as a control.

2- Two additional tubes containing 25 μls (25.625 mIU/ml) of ^{125}I-hLH only, for total concentration of hormone (CPM) computation, were set aside until counting.

3- All the assay tubes were stoppered and incubated at 25°C for 180 minutes, then the radioactivity of the bound hormone was estimated by following the steps 9, 10 and 11 in section (2.2.2).

Solutions

1- The stock solutions (0.1M) of monovalant cations were prepared by dissolving each of the amounts of salts in 250 ml ST buffer (0.01M, pH 8.2): 1.838 g KCl, 1.3375 g NH_4Cl and 1.03759 $LiCl$.

2- The stock solutions (25 mM) of divalent cations were prepared by dissolving each of the amounts of salts in 250 ml of ST buffer (0.01M, pH 8.2): 1.27 g $MgCl_2$, 1.237 g $MnCl_2$, 1.56 g $CuSO_4.5H_2O$, 0.85 g $ZnCl_2$ and 0.69 g $CaCl_2$.

3- ST buffer (0.01M, pH 8.2) was prepared as mentioned in section (2.1.7).

Calculations

1- The (SB%) was estimated as mentioned in section (2.2.4).

2- The percent of specific binding (SB%) was plotted against salts concentrations.

2.2.6 Separation of (^{125}I-hLH-Receptor) Complex by Gel-Filtration in Pre- and Postmenopausal Serous Ovarian Tumors Homogenates

Note: The experiment was carried out in duplicate.

1- One hundred microliters (150 µg protein) of benign and malignant postmenopausal ovarian tumor homogenate were added to 25 µls. (11.6 mg/ml) of ^{125}I-hLH with or without the addition of 20 fold excess of unlabeled hLH in a final volume of 0.5 ml (completed with ST buffer 0.01 M, pH 8.2). All the assay tubes were stoppered and incubated for 180 minutes at 25°C.

2- Two additional tubes containing 25 µls. (25.625 mIU/ml) of ^{125}I-hLH only, for total concentration of hormone (CPM) computation, were set a side until counting.

3- At the end of incubation the mixture was applied to the surface of a sephadex G-100 gel filtration column (1×30 cm) equilibrated with Tris buffer (0.2 M, pH 8.2). Elution was carried out using the same above buffer, to seperate ^{125}I-hLH bound to its receptors in ovarian tumor homogenate, with a flow rate of 12 ml/hr, and fraction volume of 0.5 ml.

4- The radioactivity and the absorbance at 280 nm were measured for each fraction.

5- The experiment was repeated at the optimum conditions for benign premenopausal ovarian tumor homogenate. The incubation was performed by adding 50 µls. (25.625 mIU/ml) of ^{125}I-hLH to 100 µls. (200 µg) of the protein homogenate in a final volume of 0.5 ml completed with ST buffer (0.01M, pH 7.4), then the tubes were incubated at 37°C for 120 minutes.

Calculations

1- The dimensions of the column were chosen according to the following equations[170]:

$$\text{Dimeter(cm)} = \sqrt[3]{\frac{m}{10}}$$

where

m: is the amount of protein in mg

$$\text{Length (cm)} = 30 \times \text{diameter}$$

In view of the results of such calculation, a 1×30 cm, column has been used

2- The radioactivity (CPM) and the absorbance at 280 nm were plotted against the fraction number.

Samples preparation

Benign and malignant ovarian tumors homogenates were prepared as mentioned in section (2.1.6).

Sephadex G-100 preparation

Two grams of sephadex G-100 (superfine) was allowed to swell in excess of Tris buffer 0.2 M, pH 8.2 and 7.4 (20 ml of buffer per gram of gel). During swelling excessive stirring should be avoided as it may break the beads. Then the suspension was left for 72 hours at 25°C to get equilibrated with the buffer. The buffer was decanted and the gel was resuspended in excess volume of eluent buffer three times before bed packing.

Eluent Buffer Preparation

Tris buffer (0.2M, pH 8.2 and 7.4) was prepared as follows:

Solution A: Tris 0.2M (2.4228 g), Tris (hydroxy methyl) aminomethan in 100 ml D.W.

Solution B: 0.1N HCl

Working buffer pH (8.2 and 7.4) was prepared by mixing 25 ml of solution A with an appropriate amount of solution B to adjust the pH required, then the volume was made up to 100 ml with D.W.

Bed packing

The de-gassed slurry was carefully mixed before pouring into the vertical column which contained 5 ml of the eluent buffer using a glass rod attached to the inner surface of the column. After the gel had settled the column outlet was opened. Packing was continued until the gel reached a stable bed height (30 cm). Then the column was equilibrated with Tris buffer (0.2 M, pH 8.2 and 7.4) for 24 hrs. at room temperature with dimensions of 1×30 cm and a bed volume of 20 ml.

After the gel had settled the column was sealed and then stabilized and equilibrated with eluent buffer. Revers washing of the column was also performed to get more packing of the gel in the column.

Void Volume Determination

The elution volume of Blue Dextran 2000 is equal to the column void volume (V_o), and it was determined as follows:

A fresh solution of Blue Dextran (2 mg/ml) was prepared in the eluent buffer in a sample volume of 2-3% of the total bed volume, then applied to the column with a flow rate of 12 ml/hr. Fraction of 0.5 ml were collected and their absorbance were measured at $\lambda=600$ nm.

2.2.7 The Kinetic and The Thermodynamic Studies

2.2.7.1 The Time-Course of ^{125}I-hLH Binding to Its Receptors in Pre- and Postmenopausal Patients With Serous Ovarian Tumor

Note: The experiment was carried out in duplicate.

1- At zero time, 25 and 50 μls. of ^{125}I-hLH (original conc. 25.625 mIU/ml) was added to 100 μls. (150 and 200 μg protein) of ovarian tumor homogenate. The final volume (0.5 ml) was made up by adding the assay buffer (0.01M, ST buffer pH 8.2 and 7.4). The assay tube was stoppered and incubated at 4°C for several time intervals (60, 75, 90, 105, 120, 150, 180, 240, 300 and 360 minutes).

2- Two additional tubes containing 25 and 50 μls. (25.625 mIU/ml) of ^{125}I-hLH only, for total concentration of hormone (CPM) computation, were set a side until counting.

3- After incubation, the radioactivity of bound hormone was estimated by following the steps 9, 10 and 11 in section (2.2.2).

4- Parallel experiments were performed to determine the amount of non-specific binding.

5- To determine the time-course of the association of ^{125}I-hLH with its receptor at different temperatures, the above experiment was performed at four temperatures (4, 10, 25 and 37°C).

Calculations

1- The SB% values were estimated according to section (2.2.4).

2- The percent of specific binding (SB%) was plotted against the different times of incubation at each temperature.

2.2.7.2 Determination of The Concentration of LH Receptors and The Affinity Constant of ^{125}I-hLH Association With Its Receptors in Pre- and Postmenopausal Serous Ovarian Tumors

Note: The experiment was carried out in duplicate.

1- One hundred microliters (150 and 200 µg protein) of ovarian tumor homogenate was incubated with increasing volumes (5, 10, 20, 30 and 40 µls.) of ^{125}I-hLH (0.1666-1.666 nM), with or without the addition of 20 fold excess of unlabeled hLH in a final volume of 0.5 ml (completed with ST buffer 0.01 M, pH 8.2 and 7.4). Two additional tubes containing increasing concentration (0.1666-1.666 nM) of ^{125}I-hLH only, for total concentration of the hormone (CPM) computation, were set asid until counting.

2- The assay tubes were stoppered and incubated for (180 and 120 minutes) at (25°C and 37°C) then the bound hormone was estimated by following the steps 9, 10 and 11 in section (2.2.2).

3- The previous steps were performed at different temperatures (4, 10, 25 and 37°C). The times of incubation needed to get the equilibrium state were as the following:

Temp. (°C)	Time (min)
4	300
10	240
25	180
37	120

Solutions

ST buffer (0.01 M, pH 8.2 and 7.4) prepared as mentioned in section (2.1.7).

Calculations

1- The B/F ratio was computed for each tube, where:

B: The bound radioactivity mean count (CPM), which represents the (^{125}I-hLH-receptor) complex.

F: The free radioactivity mean count (CPM) which represents the non-bound ^{125}I-hLH.

T: The total activity mean count.

F= Total count (T)- Bound radioactivity (B)

2- the value of ^{125}I-hLH bound specifically in (p mole) was calculated according to the following formula:

The value of specifically bound ^{125}I-hLH (p mole) = Specifically bound ^{125}I-hLH in p molar × incubation volume in liter

Specifically bound ^{125}I-hLH (p molar) = The percent of specific binding × Total concentration of ^{125}I-hLH in incubation medium (p molar)

$$\text{The percent of specific binding} = \left[\frac{\text{Total binding (CPM) - Non specific binding (CPM)}}{\text{Total count (CPM) of } ^{125}\text{I} - \text{hLH used in each tube}} \right] \times 100$$

3- The affinity constant and maximal binding capacity were determined according to *scatchard* equation [171].

$$\frac{B}{F} = \frac{1}{k_d} \times (B_{max} - B)$$

$$k_a = \frac{1}{k_d}$$

where:

K_a: Affinity constant

K_d: Dissociation constant

B_{max}: Maximal binding capacity

4- The plot of B/F ratios vs. the B values gives a linear relationship. The value
of the affinity constant of the binding k_a at each temperature can be
calculated from the slope of the straight line, while the value of the total
concentration of LH receptor in ovarian tumor homogenate can be calculated
from the intercept with the x-axis.

2.2.7.3 Kinetics of The Binding of ^{125}I-hLH to Its Receptors in Pre- and Postmenopausal Patients With Serous Ovarian Tumors Homogenates

Note: The experiment was carried out in duplicate.

The experiment was performed as described in section (2.2.7.1) in different
temperatures (4, 10, 25 and 37°C).

Calculations

1- The percent of specific binding (SB%) was determined according to section
(2.2.5.1) at different temperatures.

2- The rate of the association constant of (^{125}I-hLH-receptor) complex was
calculated by using the following equation:

$$\ln\left[\frac{(HR)_e}{(HR)_e - (HR)_t}\right] = K_{+1}t\left[\frac{(H)_T(R)_T}{(HR)_e}\right]$$

where:

K_{+1}:   The rate association constant

$(H)_T$:   The total molar concentration of ^{125}I-hLH

$(R)_T$:   The total molar concentration of hormone receptors.

$(HR)_e$:   The concentration of ^{125}I-hLH-receptor complex formed at
equilibrium.

$(HR)_t$:   The concentration of the complex formed after time (t).

3- The rate of the dissociation constant of the complex formed (rate of the reverse reaction constant) was calculated by using the following equation:

$$k_a = \frac{k_{+1}}{k_{-1}}$$

where

k_{-1}: The rate dissociation constant

k_a: Is the equilibrium constant of the association (affinity constant)

2.2.7.4 The Thermodynamic of ^{125}I-hLH Binding to Its Receptors in Pre- and Postmenopausal Patients With Serous Ovarian Tumors Homogenates

Note: The experiment was carried out in duplicate.

1- One hundred microliters of the protein homogenates (150 and 200 µg) were added to 25 and 50 µls. of ^{125}I-hLH in a final volume of 0.5 ml (completed with ST buffer (0.01M, pH 8.2 and 7.4). The assay tube was stoppered and incubated at 25°C and 37°C for 180 and 120 minutes.

2- After incubation, the radioactivity of (^{125}I-hLH-receptor) complex formed was estimated by following the steps 9, 10 and 11 in section (2.2.2).

3- Parallel experiments were performed to determine the amount of non-specific binding.

4- The previous steps were performed at different temperatures(4, 10, 25 and 37°C).

Calculations

1- The thermodynamic parameters of standard state were obtained from *Van't Hoff* plot, the values of the natural logarithm of equilibrium constant (affinity constant k_a) obtained at different temperatures were plotted against the reciprocal values of absolute temperature in Kelvin (1/T), according to the following equation:

$$\ln k_a = \frac{\Delta S^\circ}{R} - \frac{\Delta H^\circ}{RT}$$

where

ΔH°: The enthalpy change of the standard state.

ΔS°: The entropy change of the standard state.

R: The gas constant (8.31441 J k^{-1} mole^{-1})

ΔH° Value obtained from the slope of the linear relationship of the plot. The change in Gibbs free energy of the standard state (ΔG°) was obtained from the following equation:

$$\Delta G^\circ = -RT \ln k_a$$

While the standard state entropy change was obtained from:

$$\Delta S^\circ = \frac{\Delta H^\circ - \Delta G^\circ}{T}$$

2- The thermodynamic parameters of the transition state were obtained from **Arrhenius** plot of $\ln k_{+1}$ values against 1/T values, that gives a linear relationship according to the following equation:

$$\ln k_{+1} = \ln A - \left[\frac{Ea}{RT} \right]$$

where

A: Arrhenius constant

The value of apparent energy of activation (Ea) of the binding reaction can be determined from the slope of the straight line. The enthalpy of transition state ΔH^* obtained from:

$$\Delta H^* = Ea - RT$$

Transition state free energy change is calculated from the following equation:

$$\Delta G^* = -RT \ln k_{+1} + RT \ln\left(\frac{kT}{h}\right)$$

where k and h are *Boltzmann* and *Plank's* constants which equal (1.38×10^{-23} JK^{-1}), (0.662×10^{-33} J S^{-1}) respectively.

The change in entropy of the transition state ΔS^* is calculated from the following relation:

$$\Delta S^* = \frac{\Delta H^* - \Delta G^*}{T}$$

2.3 Spectroscopic Studies on hLH, hLH Antibody, hLH-Antibody Complex and ^{125}I-hLH-Receptor Complex

2.3.1 Factors Affecting the Absorption Properties of hLH, hLH Antibody, hLH-Antibody Complex and ^{125}I-hLH-Receptor Complex

2.3.1.1 pH Effect

• **The U.V Spectrum of hLH**

A volume of 50 µls. of hLH (200 mIU/ml) provided by (LH RIA kit from DPC-USA) was completed to 0.5 ml with different buffers at different pH values (3.8, 7.3 and 12.6). Then each of which was placed in 0.5 cm cuvette in the sample beam and the buffer at the adjusted pH in the reference beam. The absorption spectrum was measured in the area of (180-350 nm).

- **The U.V Spectrum of hLH Antibody**

A volume of 25 µls. (1.3 mg/ml) hLH Antibody provided by (LH RIA kit from DPC, USA) was completed to 0.5 ml with different buffers at different pH values (3.8, 7.3 and 12.6). Then each of which was placed in a 0.5 cm cuvette in the sample beam and the buffer at the adjusted pH in the reference beam. The absorption spectrum was measured in the area of (180-350 nm).

- **The U.V Spectrum of hLH-Antibody Complex**

1- Fifty microliters of hLH (200 mIU/ml) was added to 25 µls. of hLH antibody (1.3 mg/ml) in a final volume of 100 µls. completed with ST buffer (0.01 M, pH 7.4).

2- The assay tube was stoppered, vortexed and incubated for 120 minutes at 37°C. Then the tube was centrifuged for 30 minutes at 1500 xg in a cooling centrifuge (4°C).

3- After centrifugation the tube was placed in suitable decanting racks where the tubes were kept inverted and placed on a pad of cotton after the discard of the supernatant.

4- The precipitate was redissolved in 0.5 ml of different buffers at different pH (3.8, 7.3 and 12.6).

5- the absorption spectrum was measured in the area of (180-350 nm) using a 0.5 cm cuvette against the buffer at the adjusted pH in the reference beam.

- **The U.V spectrum of (^{125}I-hLH-Receptor)Complex in Pre-and Postmenopausal Serous Ovarian Tumors Homogenates**

1- The gel filtration experiment in section (2.2.6) gave a two peaks profile. The fractions under each peak were pooled and concentrated by dialyzing against sucrose to get the needed concentration. The absorption spectrum was measured in the area (180-350 nm), by using a 0.5 cm cuvette against Tris buffer (0.2 M, pH 8.2 and 7.4) in reference beam.

2- A volume of 50 μls. of concentrated pooled fractions under peak 1 which represents the (^{125}I-hLH-receptor) complex was completed to 0.5 ml Tris buffer (0.2 M, pH 9.0) and the absorption spectrum was measured in the area (180-350 nm) by using a 0.5 cm cuvette against Tris buffer (0.2 M, pH 9.0) in reference beam.

Solutions

1- Tris buffer (0.2 M) at different pH values (7.4, 8.2 and 9.0) was prepared as mentioned in section (2.2.6).

2- Glycin / NaOH buffer: was prepared as follows:

Solution A: Glycin (0.1 M) in NaCl (0.1 N)

(0.7507g glycin + 0.5844 g NaCl in 100 ml D.W.).

Solution B: NaOH 0.1 N

Working buffer pH (12.6) was prepared by mixing appropriate amounts of solutions A and B in a final volume of 100 ml.

3- Acetate Buffer: was prepared as follows:

Solution A: Sodium acetate 0.1 N (0.8204g $C_2H_3O_2Na$) in 100 ml D.W.

Solution B: Acetic acid 0.1 N

Working buffers pH (3.8) were prepared by mixing appropriate amounts of solutions A and B to reach the pH required in a final volume of 100 ml.

Calculations

The specific absorption coefficient a_s of hLH antibody at λ_{max}= 204.6 was calculated using **Lambert- Beer's law** [172]:

$$A = (a_s)\,(c)\,(L)$$

where:

    A: absorbance

    C: Concentration of hLH antibody in g/L

    L: Length of light path in cm

    a_s: Specific absorption coefficient in $g^{-1}.cm^{-1}.L$ at $\lambda_{max} = 204.6$ nm.

2.3.1.2 Effect of Solvent Polarity (Solvent Perturbation) on U.V Spectrum of hLH, hLH Antibody, (hLH-Antibody) Complex and (^{125}I-hLH-Receptor) Complex in Pre-and Postmenopausal Serous Ovarian Tumors Homogenates

a- The Effect of 20% of Ethanol, Glycerol, Dimethyl Sulphoxide (DMSO) and Polyethylene Glycol (PEG).

1- Fifty microliters of hLH (200 mIU/ml) was completed to 0.5 ml with Tris buffer at pH (7.3) in the presence of 20% of ethanol.

2- The mixture was placed in the sample beam using 0.5 cm cuvette against 20% ethanol prepared in the same buffer in the reference beam. The absorption spectrum was measured in the area of (180-350 nm).

3- The steps 1 and 2 were repeated to measured the absorption spectrum for hLH antibody and (hLH-antibody) complex which was prepared as described in section (2.3.1.1).

4- The effect of 20% ethanol on the U.V. spectrum of (^{125}I-hLH-receptor) complex was studied by the addition of 1 ml of Tris buffer (0.2 M) with pH (8.2 and 7.4) in the presence of 20% ethanol to 50 µls. of the concentrated pooled fractions under peak 1 which represents (^{125}I-hLH-receptor) complex as it mentioned in gel filtration experiment in section (2.2.6). The absorption spectrum was measured in the area

(180-350 nm) using a 0.5 cm cuvette against Tris buffer (0.2 M) containing 20% ethanol at the adjusted pH in the reference beam.

5- The experiment was repeated in the presence of 20% of glycerol, DMSO and PEG prepared in the same buffer.

b- Effect of Urea and KCl on the U.V Spectrum of hLH, hLH Antibody, (hLH-Antibody) Complex and (^{125}I-hLH-Receptor) Complex in Pre-and Postmenopausal Serous Ovarian Tumors Homogenates

1- Fifty microliters (200 mIU/ml) of hLH were completed to 0.5 ml using Tris buffer containing: (8 M urea, 0.03 M KCl and 8 M urea + 0.03 M KCl with a ratio of 1:1) pH (7.3). Then each of which was placed in a 0.5 cm cuvette in the sample beam against the solvent (in the same buffer) in the reference beam. The absorption spectrum was measured in the area of (180-350 nm).

2- The previous step was performed to measure the U.V. spectrum for hLH antibody and (hLH-antibody) complex which was prepared as described in section (2.3.1.1).

3- The effect of (8 M Urea and 8M Urea + 0.03M KCl with a ratio of 1:1) on the U.V. spectrum of (^{125}I-hLH-receptor) complex were studied. The complex was separated by gel filtration technique as described in section (2.2.6). A volume of 1 ml of Tris buffer (0.2M) with pH values (8.2 and 7.4) containing (8 M Urea and 8M Urea + 0.03 M KCl with a ratio of 1:1) was added to 50 μls. of concentrated pooled fractions under peak 1 which represents the (^{125}I-hLH-receptor) complex. The absorption spectrum was measured in the area (180-350 nm) by using a 0.5 cm cuvette against Tris buffer (0.2 M) containing (8M Urea and 8M Urea + 0.03 M KCl with a ratio of 1:1) at the adjusted pH in the reference beam.

Solutions

Eight molar of urea was prepared by dissolving 24.02 g of urea in 50 mls. of Tris buffer (0.2M), pH (8.2 and 7.4), 0.03 M KCl was prepared by dissolving 0.2237 g of the salt in 50 mls. of Tris buffer (0.2M, pH 8.2 and 7.4).

2.3.1.3 Observation of The Helix-Coil Transition of hLH Antibody Denaturation

1- Twenty five microliters of hLH antibody (1.3 mg/ml) were completed to 0.5 ml with a mixture of (80% H_2O+ 20% ethylene glycol) containing (0.01M) NaCl.

2- The solution was placed in 0.5 cm cuvette in the sample beam and the mixture of (80% H_2O + 20% ethylene glycol) containing (0.01M) NaCl in the reference beam.

3- The absorption was measured at the wavelength of (292 and 295 nm) at different temperatures (30, 40, 50, 60 and 70°C).

4- The experiment was repeated by using a mixture of (80% H_2O + 20% ethylene glycol) containing (0.1M) NaCl.

Solutions

Twenty percent ethylene glycol was prepared by dissolving 20 mls. of EG in 80 mls. of distilled water. 0.01M NaCl in 20% ethylene glycol was prepared by dissolving 0.05844g of NaCl in (100 mls) of 20% EG, 0.1M NaCl in 20% ethylene glycol was prepared by dissolving of 0.5844 g of NaCl in (100 mls.) of 20% ethylene glycol.

2.3.1.4 Spectrophotometric Titration of hLH Antibody

A series of hLH. antibody samples were prepared at pH range from 7.0 to 13.0 by using different buffers. The maximum absorbance of each sample was measured at wavelength of 295nm, the absorbance of λ_{max} at each pH value was plotted vs. the corresponding pH.

Another series of hLH antibody samples were prepared at pH range from 2 to 7.0 by using different buffers. The maximum absorbance of each sample was measured at wavelength of 210 nm. The absorbance of λ_{max} at each pH value was plotted vs. corresponding pH.

Preparation of the different buffers used in this experiment is described in section (2.3.1.1).

Chapter 3

Results
and
Discussion

3.1 Binding Studies of ^{125}I-Human Luteinizing Hormone With Its Receptors in Benign and Malignant Serous Ovarian Tumor

3.1.1 Tissue Collection and Processing

Three groups of patients were included in this study. Group one consisted of (8) postmenopausal patients with benign epithelial ovarian tumor (serous cystadenoma), group two contained (4) postmenopausal patients with advanced stages of epithelial ovarian cancer (serous cystadenocarcinoma) and group three contained (4) premenopausal patients with benign epithelial ovarian tumor (serous cystadenoma) as confirmed by histopathological examination. The mean age of postmenopausal patients with benign epithelial ovarian tumor was (56.125) years and the age ranged from (54-58) years, while the mean age of postmenopausal patients with epithelial ovarian cancer was (57.16) years and the age ranged from (54-59) years. The mean age of premenopausal patients with benign epithelial ovarian tumors was (42.0) years and the age ranged from (40-44) years.

The weights of resected tissue samples ranged between (0.5-2.5)grams. Tissue homogenization was carried out in 0.25M sucrose. Sucrose is a hypotonic solution that enhances the rupture of plasma cell membrane and preserves other cell organelles [173].

The homogenization of the samples was carried out in a cold medium (4°C), to avoid protein denaturation and to decrease the proteolytic enzymes activity which might damage the hormone receptors, so low temperature is a part of avoiding proteolysis[137,174]. The tissue homogenate was filtered through several layers of nylon gauze to remove connective tissue fragments and debris, while centrifugation at 1000 xg removed the unruptured cells and intact nuclei of the ruptured cells, leaving mitochondrial/golgi fractions and cell microsomes in the supernatant[175]. Ovarian homogenate containing the mitochondrial/golgi fractions was used as the receptor source in our study.

3.1.2 Determination of ^{125}I-hLH Concentration

The concentration of ^{125}I-hLH (mIU/ml) was determined according to self-displacement method suggested by *Morris*[169] for the assessment of the concentration of any radioactive tracer. This method uses increasing amounts of ^{125}I-hLH for the incubation with constant amount of LH antiserum under conditions similar to those employed in normal radioimmunoassay (RIA) for which the tracer is used. Specific radioactivity and the tracer concentration were then determined by comparing the ratio of bound to free (B/F) at each increment of the tracer with (B/F) for the normal RIA standard curve. Self displacement method necessitates the presence of standard solutions of labeled with concentrations higher than that used in the normal RIA. The final volume of all solutions used in both two parallel experiments must be equal. Due to our limited facilities and unavailability of standard solutions of ^{125}I-hLH with concentrations higher than that present in LH RIA Kit, the method was modified by adding different but comparable volumes of ^{125}I-hLH (i.e., 150, 200, 250, 300 and 350 μl) that caused a different concentration of the ^{125}I-hLH in the incubation medium. According to this method the approximate estimation of the concentration of ^{125}I-hLH was found to be (25.625 mIU/ml).

3.1.3 Determination of Human Luteinizing Hormone Levels in Sera of Postmenopausal Serous Ovarian Tumor Patients and Controls

Serum hLH levels were measured with radioimmunoassay (RIA) in two groups of postmenopausal patients with epithelial ovarian tumor matched with one group of control subject. Group 1 contained (8) postmenopausal patients with benign epithelial ovarian tumor (serous cystadenoma), Group II comprised of (4) postmenopausal patients with epithelial ovarian cancer (serous cystadeno-carcinoma). Table (3-1) and figure (3-1) show the results obtained from this study. The level of serum hLH in postmenopausal patients with benign tumor was found to be (30.84 mIU/ml), whereas that of postmenopausal patients with malignant tumor was found to be (28.6 mIU/ml). But in controls, the level was found to be (33.26 mIU/ml). Student's T-test analysis revealed that there is no significant decrease of serum hLH levels (P>0.1), in benign and malignant postmenopausal serous ovarian tumors patients.

Blaakaer et. al., reported that there were no significant differences found in serum hLH levels in postmenopausal women with benign or malignant epithelial ovarian tumors, while significant lower serum FSH levels were demonstrated in women with malignant tumors[176]. *Jeppsson et. al.,* found that women with epithelial ovarian tumors had lower serum levels of FSH and LH than was normal [177].

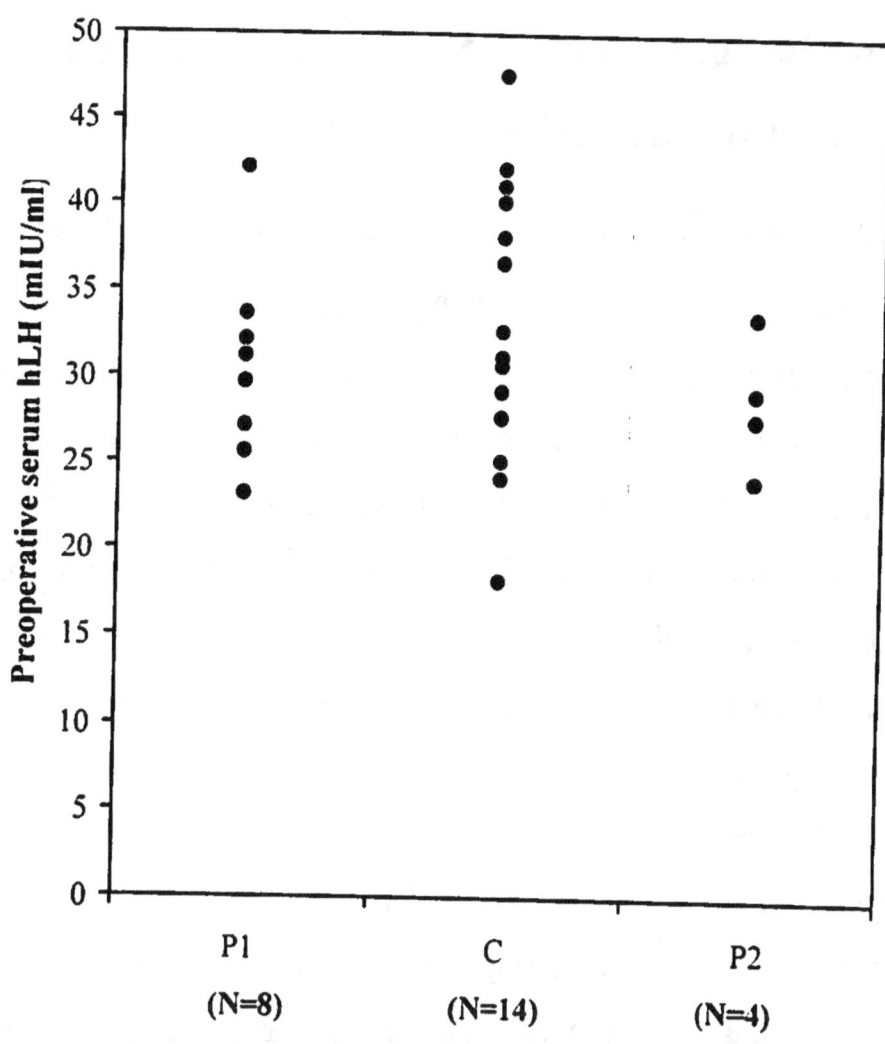

Figure(3-1): Distribution of individual pre-operative serum hLH values (mIU/ml) in:
- Postmenopausal of benign serous ovarian tumor (P₁)
- Postmenopausal of serous ovarian cancer (P₂)
- postmenopausal normal women (C)

Details are described in section (2.2.3)

Table (3-1): Pre operative serum hLH concentrations (mIU/ml) in patients with serous ovarian tumors. Details are described in section (2.2.3)

Group	No. of cases	Age (year)	Serum hLH (mIU/ml)
Postmenopausal of serous ovarian cancer	4	57.16 ± 1.456	28.6 ± 3.98
Postmenopausal of benign serous ovarian tumor	8	56.25 ± 2.06	30.84 ± 5.8
Control	14	56.77 ± 1.8	33.26 ± 8.105

3.2 Radio Receptor Assay Studies of ^{125}I-hLH Binding to Its Receptors in Serous Ovarian Tumor

3.2.1 Preliminary Test of ^{125}I-hLH Binding to Its Receptors in Serous Ovarian Tumor Homogenate

Luteinizing hormone receptors, which are located at the cell membrane, were investigated in 16 cases of benign and malignant serous ovarian tumors. Membrane hLH receptors were detected through the incubation of ^{125}I-hLH with ovarian tumor homogenate and the bound (^{125}I-hLH) was measured by radioreceptor assay method.

Preliminary experimental conditions used resulted in 18% specific binding in the postmenopausal serous ovarian cancer patients, 13% specific binding in postmenopausal patients with benign serous ovarian tumor and 11% specific binding in premenopausal patients with benign serous ovarian tumor.

The data obtained, in this study revealed also that the tumors of serous ovarian cancer patients had higher incidence of hLH receptors than those of benign groups, and the benign tumor of postmenopausal patients included higher incidence of hLH receptors than those of premenopausal patients.

3.2.2 Most Appropriate Conditions of the Binding of ^{125}I-hLH to Its Receptors in Serous Ovarian Tumor

3.2.2.1 Effect of hLH Receptor Concentration on ^{125}I-hLH Binding With Its Receptors in Serous Ovarian Tumor Homogenate

To determine whether the specific binding was proportional to the amount of receptors protein actually present in the incubation mixture, increasing amounts of protein in the homogenate were incubated with either tracer ^{125}I-hLH alone, or with unlabeled hormone added.

Figure (3-2) shows that the specific binding of ^{125}I-hLH to its receptors in ovarian tumors was easily detected at the lowest particulate concentration tested (equivalent to 25 µg/ml protein), and increased with increasing amount of the homogenate added to the incubation medium. Two hundred microgram of benign ovarian tumor homogenate protein for premenopausal patients and 150 µgs. of benign and malignant ovarian tumor homogenate protein for postmenopausal patients were used in all the subsequent experiments since it gives maximum value of the specific binding.

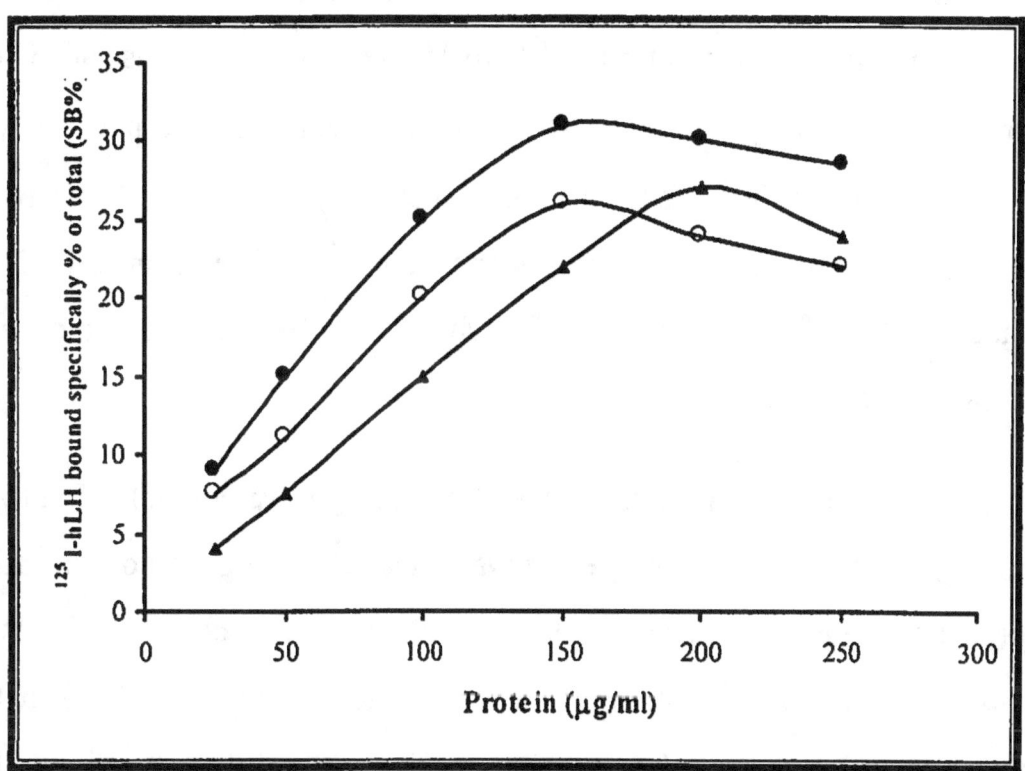

Figure (3-2): Influence of receptor concentration on the binding of ^{125}I-hLH with,

    ▲   benign premenopausal ovarian tissue homogenate

    O   benign postmenopausal ovarian tissue homogenate

    ●   postmenopausal ovarian cancer tissue homogenate

Details are described in section (2.2.5.1)

3.2.2.2 Effect of ^{125}I-hLH Concentration on The Binding With Its Receptors in Serous Ovarian Tumor Homogenate.

Gonadotrophin (Gn) uptake by a tissue must meet certain criteria before it can be considered to represent a hormone-receptor interaction.

In particular, there must be a finite number of high affinity sites per unit tissue. Consequently, gonadotrophin binding will be a saturable process, such that specific uptake per unit tissue will not increase beyond a maximal level despite further addition of gonadotrophin[178].To fulfil this criterion and to estimate the suitable concentration of ^{125}I-hLH, the experiment was carried out in the presence of 200μg of receptor protein for premenopausal patients with benign tumor, and 150μg for postmenopausal patients with benign and malignant tumors, and increasing concentration of ^{125}I-hLH. Figure (3-3) is representative of ^{125}I-hLH binding curve with its protein receptors in ovarian tumor homogenate.

As shown in the same figure, the specific binding of radiolabeled LH to its receptors in ovarian tumor homogenate was linear at low ligand concentration and approached a plateau at a concentration that was equivalent to (1.666 nM, 50μl) in premenopausal patients with benign tumor and (0.833 nM, 25μl) in postmenopausal patients with benign and malignant tumors. According to the results of this experiment (1.666 nM and 0.833nM) of tracer hLH were used in the binding studies in the subsequent experiments.

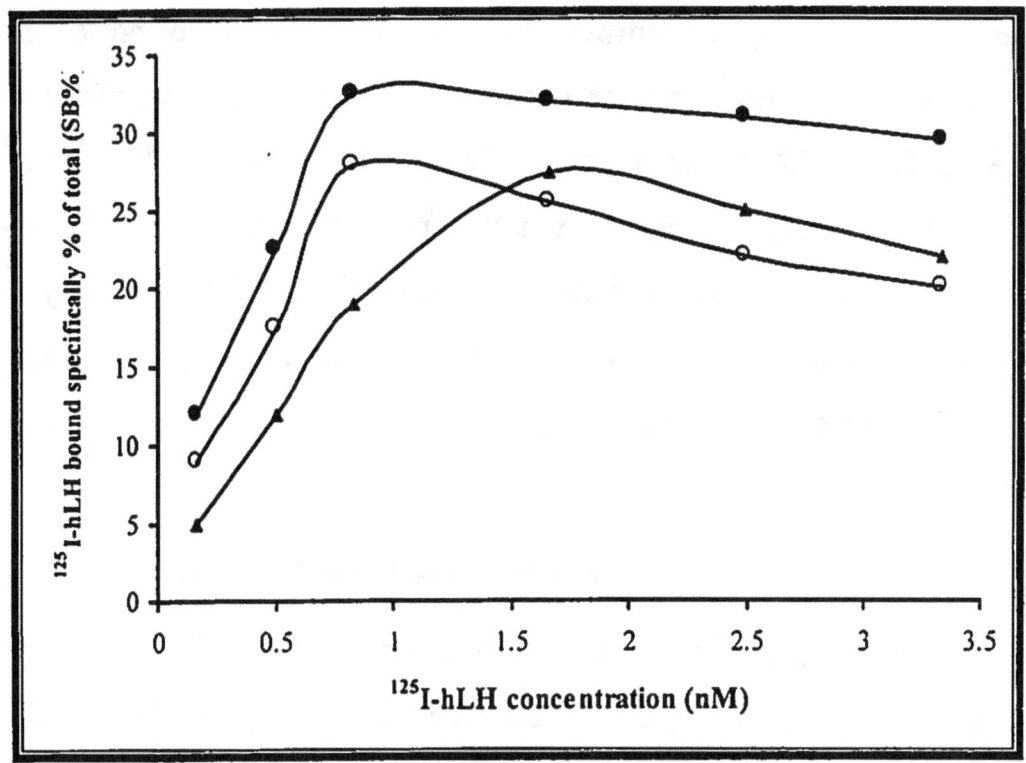

Figure (3-3): Effect of different concentration of ^{125}I-hLH on the binding with its receptor in:

    ▲   premenopausal patients with benign tumor

    ○   postmenopausal patients with benign tumor

    ●   postmenopausal patients with malignant tumor

Details are described in section (2.2.5.2)

3.2.2.3 Effect of Different pH on ^{125}I-hLH Binding With Its Receptors in Serous Ovarian Tumor Homogenate

The analysis of the influence of pH on binding of ^{125}I-hLH to its receptors in ovarian tumor homogenate is stated in figure (3-4). It can be seen from figure(3-4) that there is a broad pH maximum for binding and the apparent maximum occurs at a pH value higher than physiological (pH 8.2) for benign and malignant postmenopausal ovarian tumors, while pH 7.4 is the optimum pH for the binding of ^{125}I-hLH to the benign premenopausal ovarian tumor. These results indicate that the binding is pH dependent and the shift in the pH of the environment may affect the properties of the macromolecules involved in the binding. This effect includes the induction of protonation-deprotonation processes occurring with the ionizable groups of the amino acids present in the binding domain of these macromolecules[179]. The maximal binding more favorable at pH(8.2, 7.4) than the low pH, this may be due to acid base hormone binding properties of each site or possible protein conformational changes induced by changes in the pH and thus account for the functional heterogeneity of the binding sites[180]. These results could be explained by assuming the existence of two inter-convertible states of the receptor-hormone complex to the following equation:

$$\text{LH + Receptor} \rightleftharpoons \text{LH - Receptor} \rightleftharpoons \text{[LH-Receptor]}^- + \text{H}^+$$

where the hormone (LH) interacts with the receptor (R) to form the complex [LH-R]. The complex undergoes a slow transition into a second state [LH-R]$^-$ upon release of proton (H$^+$). According to this model, hormone binding as conventionally measured is apparently slow, since it represents mainly the rate of transition of the stable state [LH-R]$^-$ whereas [LH-R] rapidly dissociates to R and LH upon dilution or washing of the membranes. This model predicts that cumulation of the [LH-R]$^-$ will not be favored at low pH, consistent with the low binding observed at pH 5.4. Moreover, receptor-hormone complex formed at pH 8.2 and 7.4 should rapidly dissociate upon acidification, and indicate that the release of the hormone is reversible and was not resulted from destruction of the receptor[181].

Extensive studies indicated that there were different effects of pH on the binding of ^{125}I-hLH to its receptors. *Hannu et. al.,* reported that the highest binding of radiolabeled hLH to luteal homogenate from human ovarian corpora lutea was seen at pH 7.0-7.5 [182]. *Cameron et. al.,* reported that the maximum binding of ^{125}I-hLH to luteal membranes from the Rhesus monkey was between pH 6.0-7.5 and diminished sharply above and below this range[183], while *Yehudith et. al.,* reported that pH 7.0 is the optimum pH for the binding of ^{125}I-hLH to purified plasma membranes from rat ovary[181]. The difference in the pH range suggests that difference in target tissues may posses different binding domains for the labeled hormone and may be due to the difference in binding conditions.

According to the results obtained in this experiment, the pH of incubation buffer in all subsequent experiments was adjusted at (7.4 and 8.2) as optimum pH for the three different groups used in this study.

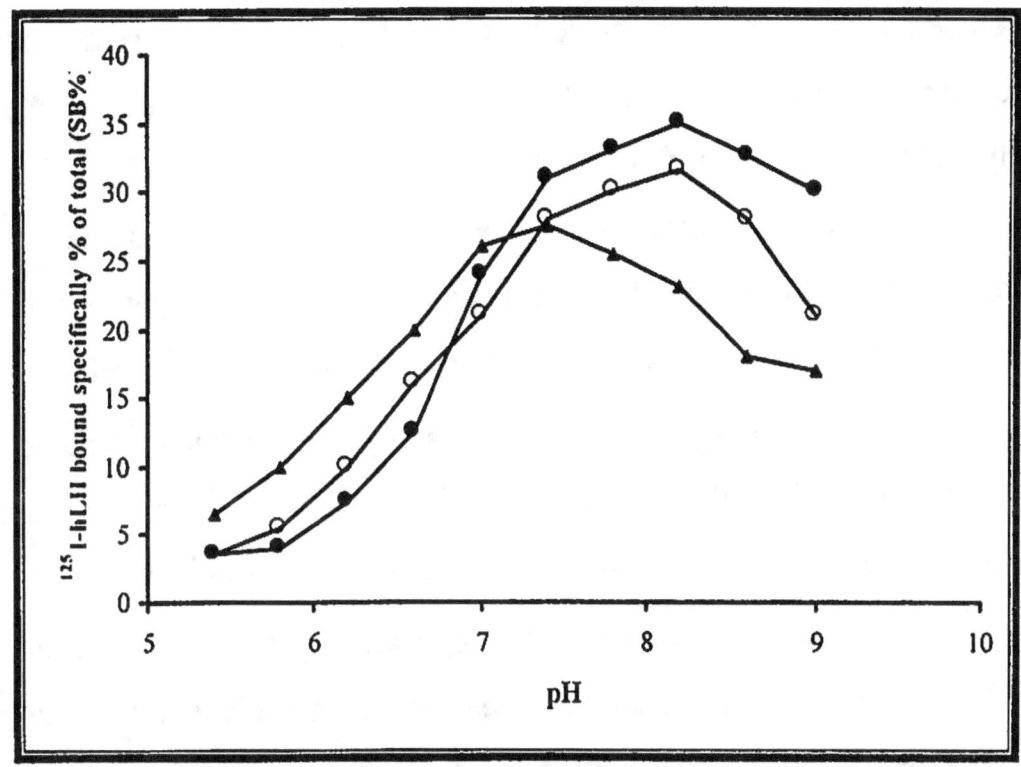

Figure (3-4): Effect of pH on the binding of ^{125}I-hLH to its receptor in:
 ▲ premenopausal patients with benign tumor
 O postmenopausal patients with benign tumor
 ● postmenopausal patients with malignant tumor
Details are described in section (2.2.5.3)

3.2.2.4 Effect of Temperature on the Binding of ^{125}I-hLH to Its Receptors in Serous Ovarian Tumor Homogenate

The temperature dependency of the binding of ^{125}I-hLH with its receptor in both benign and malignant ovarian tumors was examined. Figure (3-5) shows the results of this analysis. It seemed that specific binding was increased when the temperature was raised and a maximum binding obtained at 25°C for benign and malignant postmenopausal ovarian tumors and 37°C for benign premenopausal ovarian tumor. A decrease in the specific binding at 45°C may be due to the fact that dissociation rates show a greater increase with temperature than association rates, giving a higher affinity constant at lower temperature, or it's may be due to denaturation of receptor molecules which occurs with increasing temperatures [137].

Temperature is known to affect protein folding, so previous study suggests that decreased temperatures can allow partially misfolded LH receptors to fold properly and be expressed in functional forms[184].

Our results are consistent with those obtained by others who reported that the binding was temperature dependent and that the optimum temperature for the binding of hLH to its receptors is 22°C in ovarian rat homogenate[182], 25°C in granulosa cells isolated from rat ovary[185] and 25°C in human ovarian cancer homogenate[186]. While others reported that ^{125}I-hLH binding experiments were performed at 37°C in pseudo pregnant rat ovarian extract [187] and in human benign and malignant ovarian tumor homogenate [188]. The difference in the appropriate incubation temperature for the binding may be due to the histologic tumor type or may be due to the fact that the fluidity and conformation of cell membranes under these different temperatures will not be the same [137], or to the different binding conditions.

As a result of the temperature sensitivity of hLH receptor complexes it was adviced to study the time course of the association of the ^{125}I-hLH with their receptors at different temperatures.

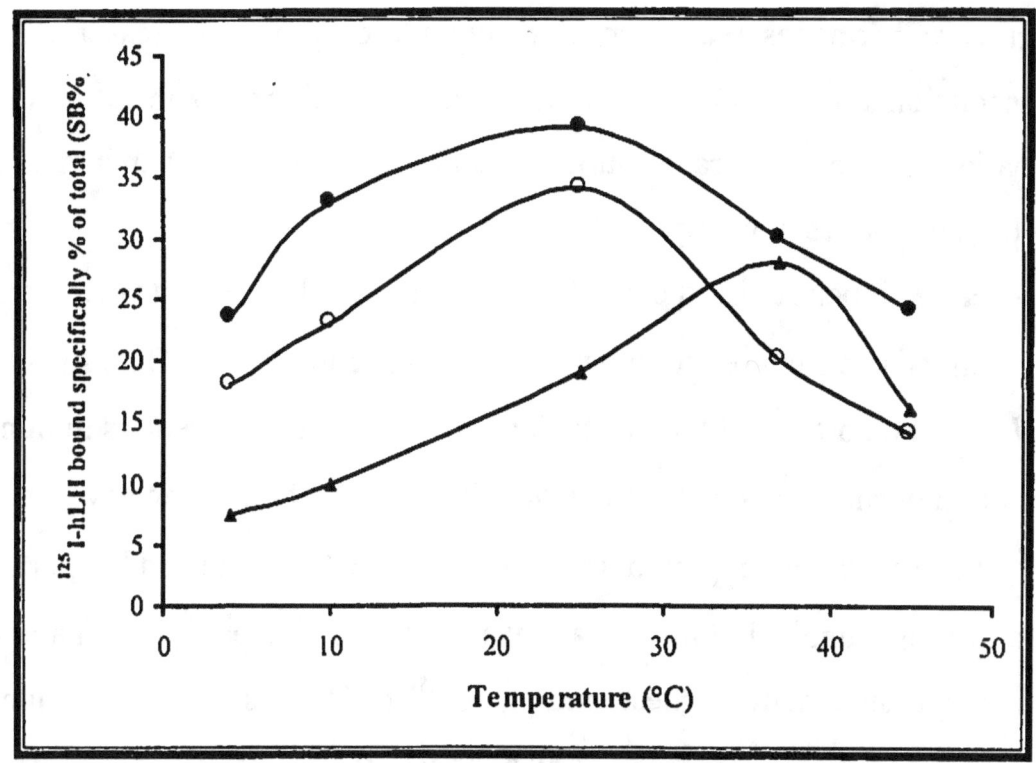

Figure (3-5): Effect of the temperature on the binding of ^{125}I-hLH with its receptors in:

    ▲ premenopausal patients with benign tumor

    ○ postmenopausal patients with benign tumor

    ● postmenopausal patients with malignant tumor

Details are described in section (2.2.5.4)

3.2.2.5 The Choice of Most Appropriate Incubation Time for The Binding of ^{125}I-hLH With Its Receptors in Serous Ovarian Tumors Homogenate

To choose the most appropriate incubation time at 25 and 37°C, the experiment was carried out at different time intervals (60-360 minute).

Figure (3-6) shows that the optimal binding of ^{125}I-hLH to its receptors in premenopausal patients with benign tumor was occurred within 120 min. incubation at 37°C, and in postmenopausal patients with benign and malignant tumors, the maximum specific binding was occurred within 180 min. incubation at 25°C.

Hormone interaction with receptors must be a very rapid process in vivo, while in vitro hormones react very slowly with receptors. The reason for this slow association rate is not clear, but it may depend on an alteration in the conformation of the membrane during isolation, since receptors on intact cells usually bind hormones more rapidly [137].

Lee et. al., reported that the maximal binding of ^{125}I-hLH to its receptors in ovarian homogenate from pseudopregnant rat was occurred at 25°C for 4hrs.[189], *Richard et. al.,* reported that the incubation of ^{125}I-hLH to its receptors in human ovarian homogenate was performed at 25°C for 20 hrs. to achieve steady state[186], *Hannu et. al.,* reported that 19 hrs. at 22°C, a time and temperature sufficient for maximal binding in rat ovarian homogenate[190], and in another study they reported that the highest binding of ^{125}I-hLH to its receptors in human corpora lutea homogenate was seen at 37°C for 60 min.[182]. While *Lee et. al.,* reported that the total binding of ^{125}I-hLH to its receptor in human corpora lutea homogenate was approached a plateau at 30 min. at 37°C[191]. Maximum specific binding at different incubation time were also reported by other authors, this could be due to the different source of LH receptors [181, 183, 188].

According to our results, the binding studies of the subsequent investigations were carried out at 37°C for 120 min. incubation for the premenopausal group, and at 25°C for 180 min. incubation for the postmenopausal groups.

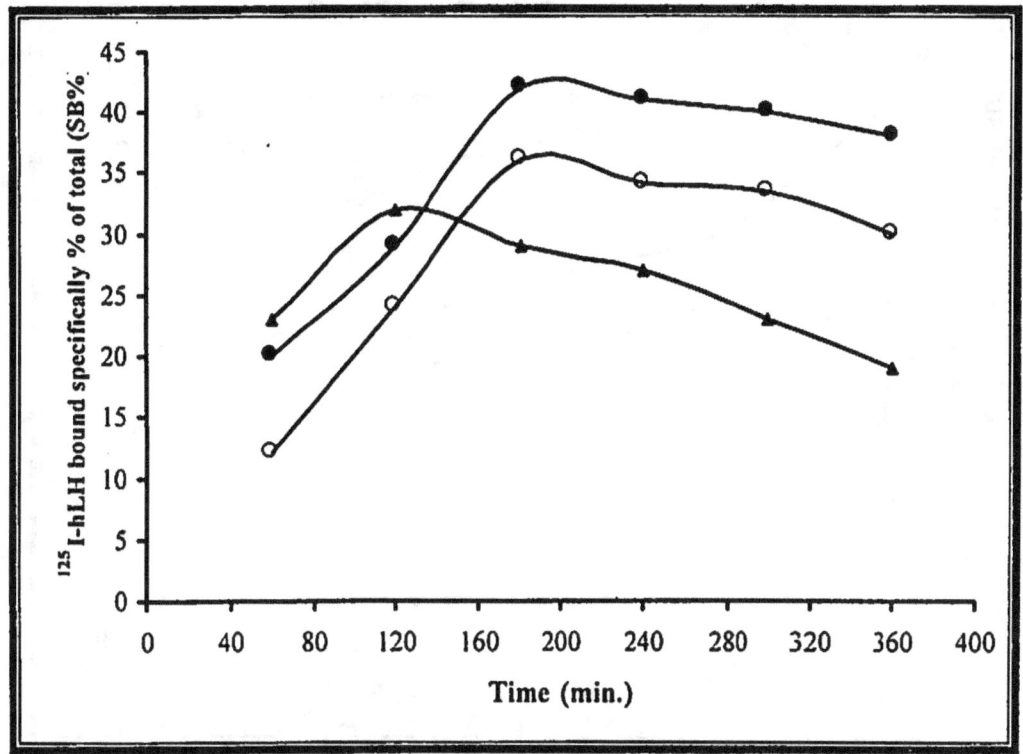

Figure (3-6): Time-course of ^{125}I-hLH binding with its receptors in:

▲ premenopausal patients with benign tumor

○ postmenopausal patients with benign tumor

● postmenopausal patients with malignant tumor

Details are described in section (2.2.5.5)

3.2.2.6 Stability of ^{125}I-hLH-Receptor Complex of Postmeno-pausal Serous Ovarian Tumor

The influence of temperature on the stability of ^{125}I-hLH-receptor complex as a function of time was studied. The complexes of benign and malignant ovarian receptors were reincubated at four temperatures (4, 25, 37 and 45°C) and at certain time intervals the remaining bound LH was estimated. As seen in figure (3-7A & B), the rate of dissociation of ^{125}I-hLH-receptor complex was increased as the temperature increased. At 4°C the complex remained stable for 8 hrs. While it dissociated weakly at 25°C after the same period. When increasing the temperature to 37°C and after 8 hrs about 48% and 32% of the complex were dissociated in benign and malignant tumors respectively. But further increase in temperature to 45°C caused 70% and 60% dissociation.

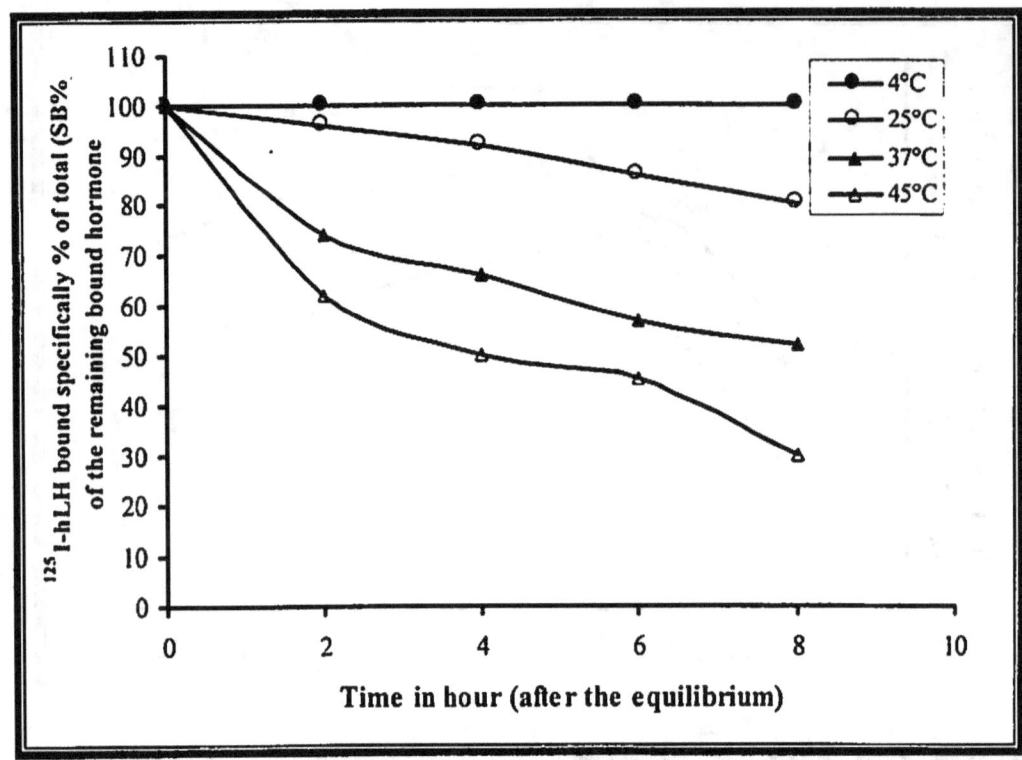

(a)

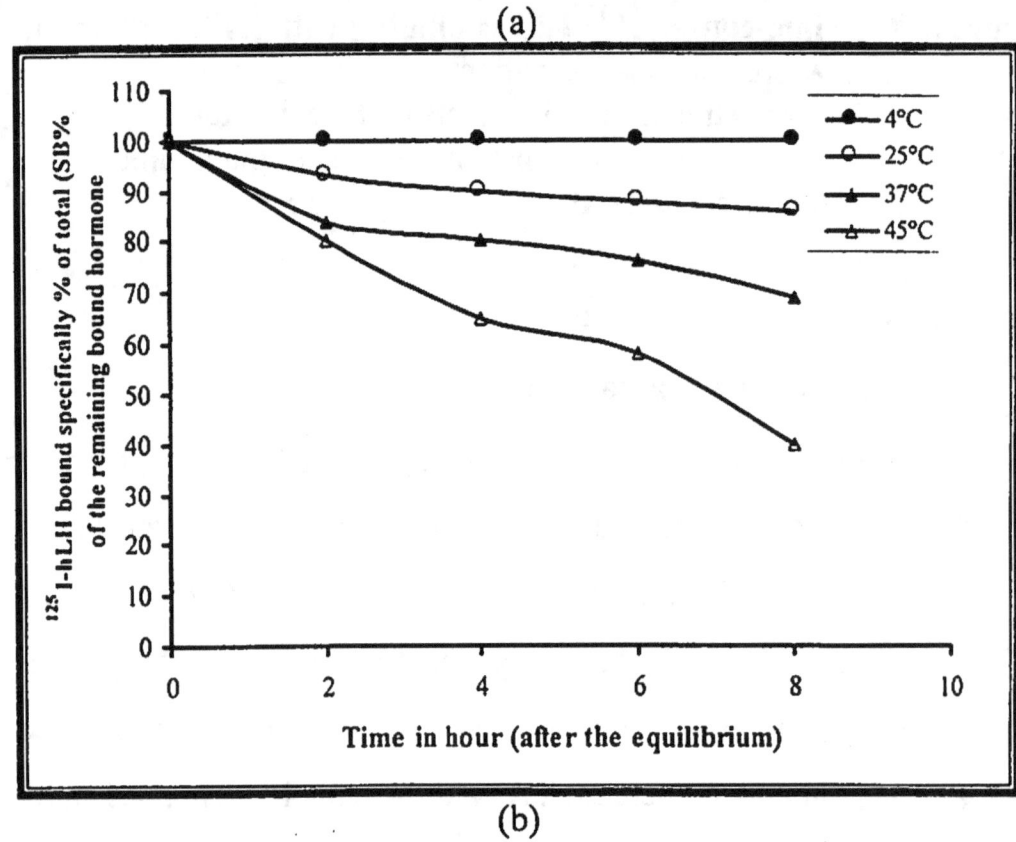

(b)

Figure (3-7): Stability of (^{125}I-hLH-receptor) complex of:
(a) Postmenopausal patients with benign tumor
(b) Postmenopausal patients with malignant tumor, at four different temperatures.
Details are described in section (2.2.5.6)

3.2.2.7 Competitive Effect of Different Concentrations of Different Unlabeled Hormones on the Binding of ^{125}I-hLH to Its Receptors in Postmenopausal Serous Ovarian Tumor Homogenate

The specificity of ^{125}I-hLH binding was established by determining the ability of different concentrations of glycoprotein hormones to compete with the radioligand for the binding sites. The inhibition of the binding of radiolabeled LH to its receptors in postmenopausal ovarian tumors homogenate by unlabeled LH, hCG, FSH and TSH is illustrated in figure (3-8 A & B). Unlabeled LH and hCG inhibited the binding in a dose-related manner. A significant inhibition occurred at a concentration of (60mIU/ml) of either hormone, unlabeled hCG appeared to be slightly less potent than LH; TSH did not inhibit binding and a small effect was also obtained with FSH at a concentration of (90mIU/ml). The displacement of ^{125}I-hLH by hCG related to its structure that shares tight homologies with the binding domains of hLH, also the addition of unlabeled hormone would cause a redistribution of bound tracer from high affinity; slow dissociating sites, to low affinity; faster dissociating sites[137]. Several studies have illustrated the competitive effect of unlabeled hormones on the binding of ^{125}I-hLH to its receptors in different gonads. *Kolena et. al.* reported that binding of ^{125}I-hLH to porcine follicular fluid is inhibited by unlabeled hLH, hCG but not FSH, ACTH and GH [192], *Jia et. al.* reported that hCG and hLH displaced ^{125}I-hLH binding to rat ovarian homogenate, incontrast FSH was not effective[193]. Other data also reported that bound ^{125}I-hLH was not displaced by FSH and TSH [130].

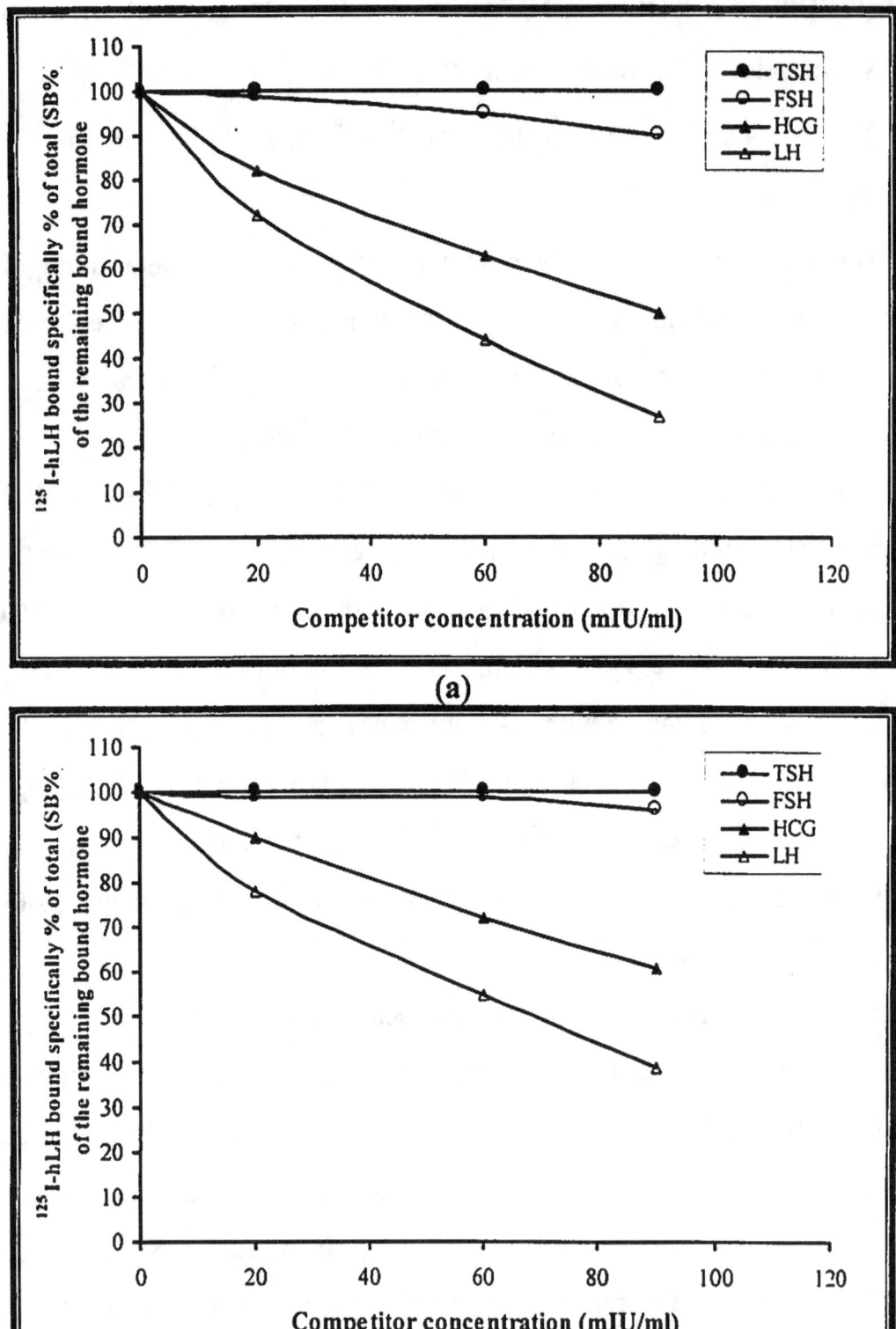

Figure (3-8): Binding of, (a) benign postmenopausal receptors (b) malignant postmenopausal receptors, with ^{125}I-hLH in the presence of different concentrations of unlabeled competitors.

Details are described in section (2.2.5.7)

3.2.2.8 Effect of Different Halides on ^{125}I-hLH Binding With Its Receptor in Postmenopausal Serous Ovarian Tumor Homogenate

Figure (3-9) shows the effect of different halides salts (i.e. NaF, NaCl, NaBr, and NaI) at 0.1M concentration on the extent of binding of ^{125}I-hLH to its receptors in ovarian tumor homogenate. It seemed that the sodium halides inhibited the ^{125}I-hLH-receptor binding according to the following order:

$$NaI > NaBr > NaCl > NaF$$

The order corresponds to the decreasing ionic radius and increasing radius of hydration. Presumably, the lesser degree of hydration permits greater interaction of the salt with an ionic group located in the hormone or receptor combining site[148].

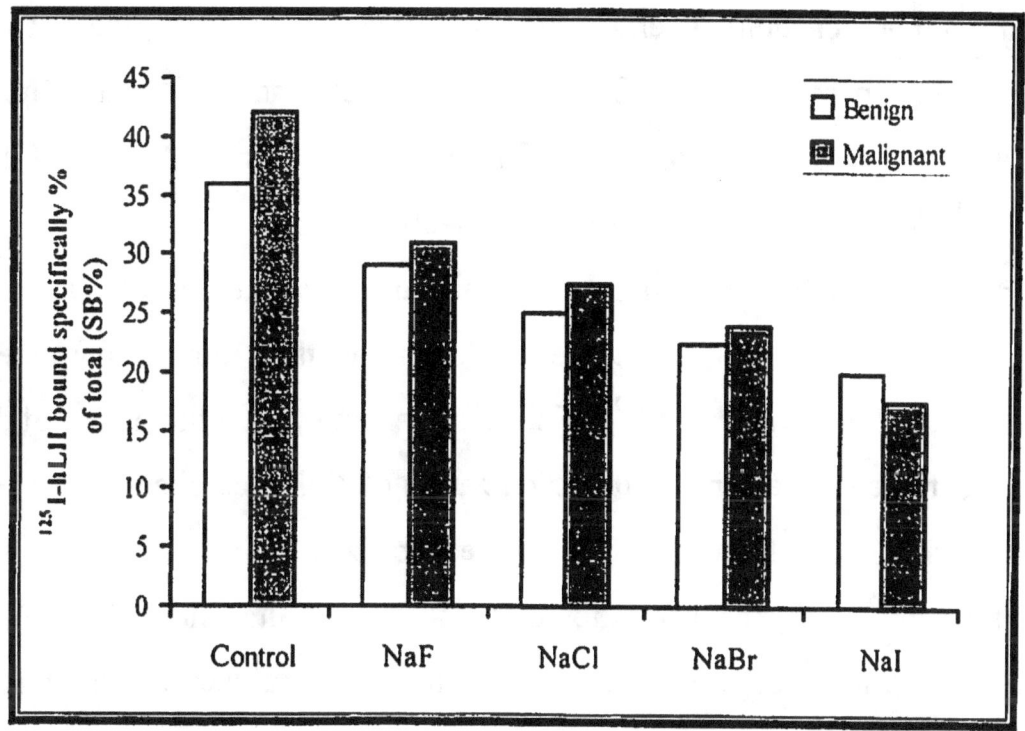

Figure (3-9): Effect of different halides on the binding of ^{125}I-hLH to its receptors in benign and malignant postmenopausal ovarian tumors.
Details are described in section (2.2.5.8).

3.2.2.9 Effect of Monovalent And Divalent Cations on the Binding of ^{125}I-hLH With Its Receptor in Postmenopausal Serous Ovarian Tumors Homogenate

The importance of the ionic environment for the binding of ^{125}I-hLH to its receptors in ovarian tumors is illustrated in figure (3-10). Monovalent cations appeared to inhibit the reaction or dissociate the (hormone-receptor) complex. These results are in agreement with those obtained by other authors, who reported that the order of inhibitory potency of monovalent cations was LiCl>NaCl>KCl[194]. The effect of NH$_4$Cl was examined and it seemed to inhibit the ^{125}I-hLH binding to its ovarian receptors, **Ralph et. al.** reported that NH$_4$Cl decreased the amount of ^{125}I-hLH binding activity [195].

Some of divalent cations appeared to enhance the binding reaction, while others inhibit the reaction. Several cations have been used to study their action on the binding, these cations are Ca^{+2}, Mg^{+2}, Mn^{+2}, Cu^{+2} and Zn^{+2} in a (25mM) concentration. The presence of Mg^{+2}, Ca^{+2} and Zn^{+2} seemed to inhibit the binding of ^{125}I-hLH to its ovarian receptors, while presence of Mn^{+2} seemed to increase the specific binding and Cu^{+2} also increased the binding but to a lesser extent. This is presumably due to their effect to activate the catalytic component in the hormone or in the receptors[194]. The metal ions may alter the nature of the hydrophobic forces necessary for the stabilization of biological membranes and affect the hydrophobic forces controlling the stabilization of ^{125}I-hLH receptor complex formed[196]. High concentration of metal ions or salts tend to destabilize membrane as results of their interaction with water molecules leading to diminishing of protein-protein interaction and reversible denaturation of the receptor[197].

Several studies demonstrated the effect of different divalent cations on the binding of ^{125}I-hLH to its receptors in tissue and target cells. **Hennu et. al.** reported that considerable reduction in binding was observed at a concentration between (10-100 mM) of $MgCl_2$ or $CaCl_2$[182], and another study indicated that $MgCl_2$ method dissociated bound hormone from LH receptor sites without any alteration of the binding sites[198]. The inhibiting effect of $ZnCl_2$ on the ^{125}I-hLH binding to its receptors may be due to that zinc is capable of binding with specific sites on receptor molecule and then inhibiting the hormone binding[199,200].

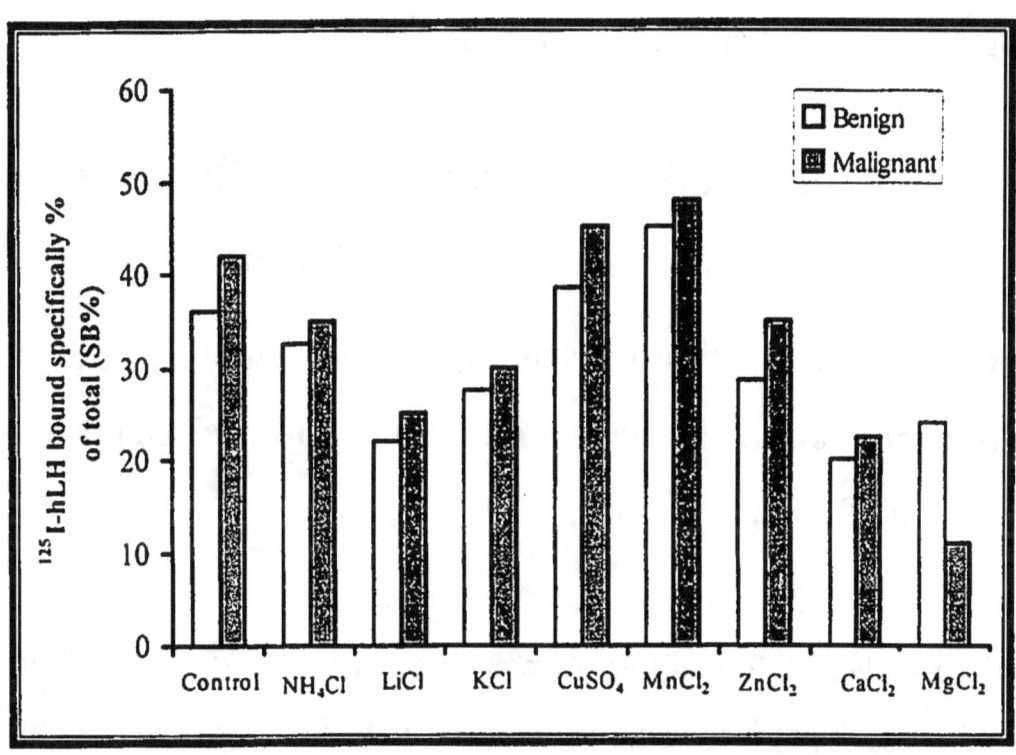

Figure (3-10): **Effect of monovalent and divalent cations on the binding of ^{125}I-hLH with its benign and malignant postmenopausal ovarian receptors.**
Details are described in section (2.2.5.9).

3.2.3 Separation of (^{125}I-hLH-Receptor)Complex by Gel-Filtration in Serous Ovarian Tumor Homogenate

Figure (3-11 B,C & D) shows the results of gel filtration elution patterns of ^{125}I-hLH binding with its receptors in ovarian tumor homogenate on a column of sephadex G-100 (1×30 cm) equilibrated with 0.2M Tris buffer (pH 7.4, 8.2).

The void volume of this column was 4ml as predicted from the elution profile of the blue dextran as shown in figure (3-11 A). The radioactivity was distributed in the gel filtrate in two peaks, the first peak represented (hormone-receptor) complex and the second peak was corresponding to the unbound (Free) ^{125}I-hLH. The fractions under each peak were pooled, concentrated and then subjected to spectroscopic studies as mentioned in section (2.3).

The elution profiles resulted from our experiment were consistent with the same observations found in a number of studies [182, 189, 201].

3.3 The Kinetics and Thermodynamics of the Interaction of Luteinizing Hormone With Its Receptors

3.3.1 Kinetics of the ^{125}I-hLH Binding to Its Receptors in Pre-and Postmenopausal Patients with Benign and Malignant Serous Ovarian Tumors

Figure (3-12 A,B & C) shows the time-course of the formation of (^{125}I-hLH -receptor) complex at four different temperature (4, 10, 25 & 37°C) in two groups of pre-and postmenopausal patients with benign serous ovarian tumors and one group of postmenopausal patients with serous ovarian cancer.

The results of time course patterns at different temperatures revealed that the binding of ^{125}I-hLH to its receptors in serous ovarian tumors is a temperature and time dependent process with a maximum binding occurs at 37°C with 120min for premenopasual patients with benign tumor and at 25°C with 180 min for postmenopausal patients with benign and malignant ovarian tumors.

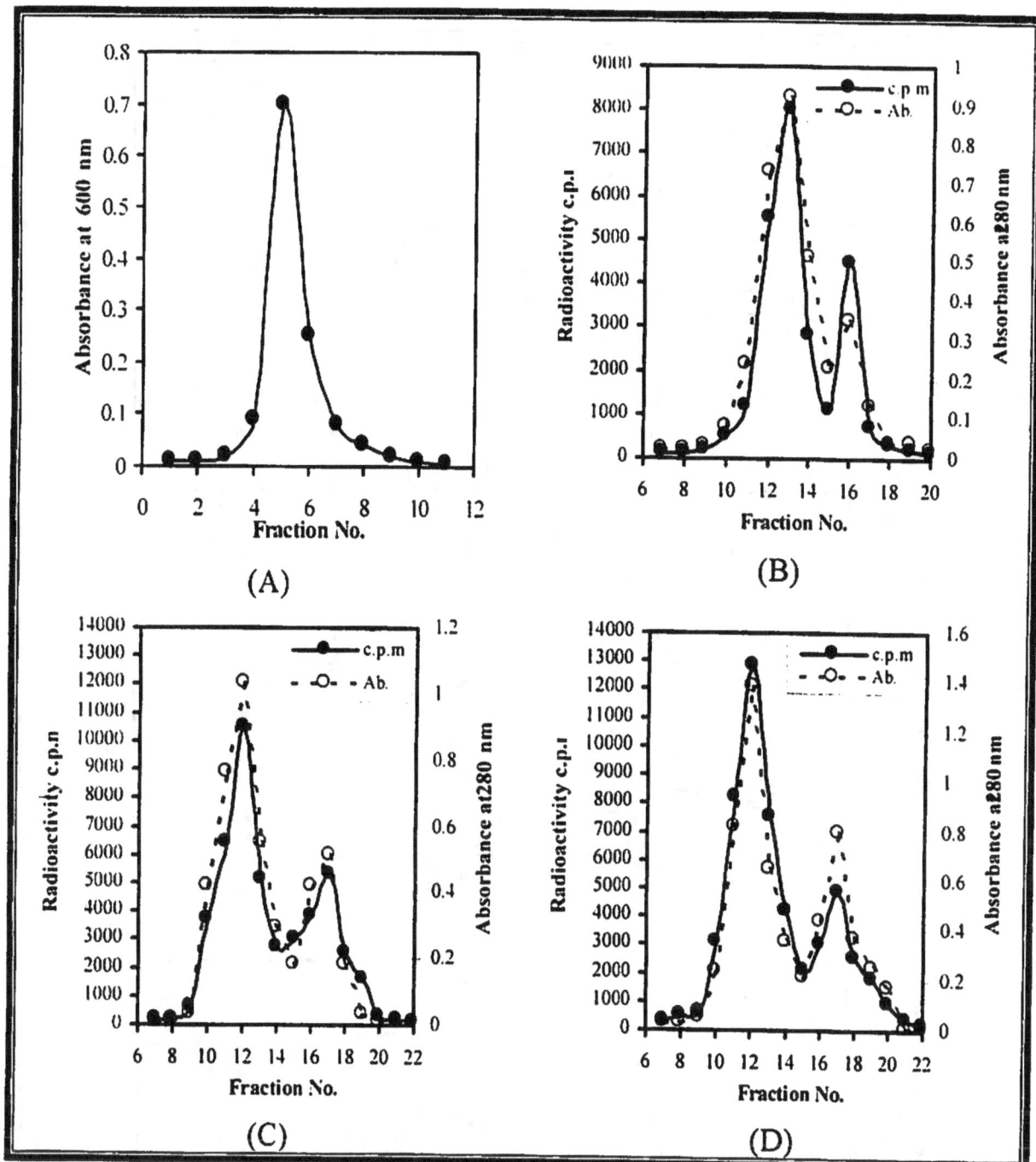

Figure (3-11): Elution profile of:

    A-   Blue Dextran 2000,

    B- ^{125}I-hLH-receptor complex from premenopausal patients with benign ovarian tumor,

    C- ^{125}I-hLH-receptor complex from postmenopausal patients with benign ovarian tumor,

    D- ^{125}I-hLH-receptor complex from postmenopausal patients with ovarian cancer.

Details are described in section (2.2.6)

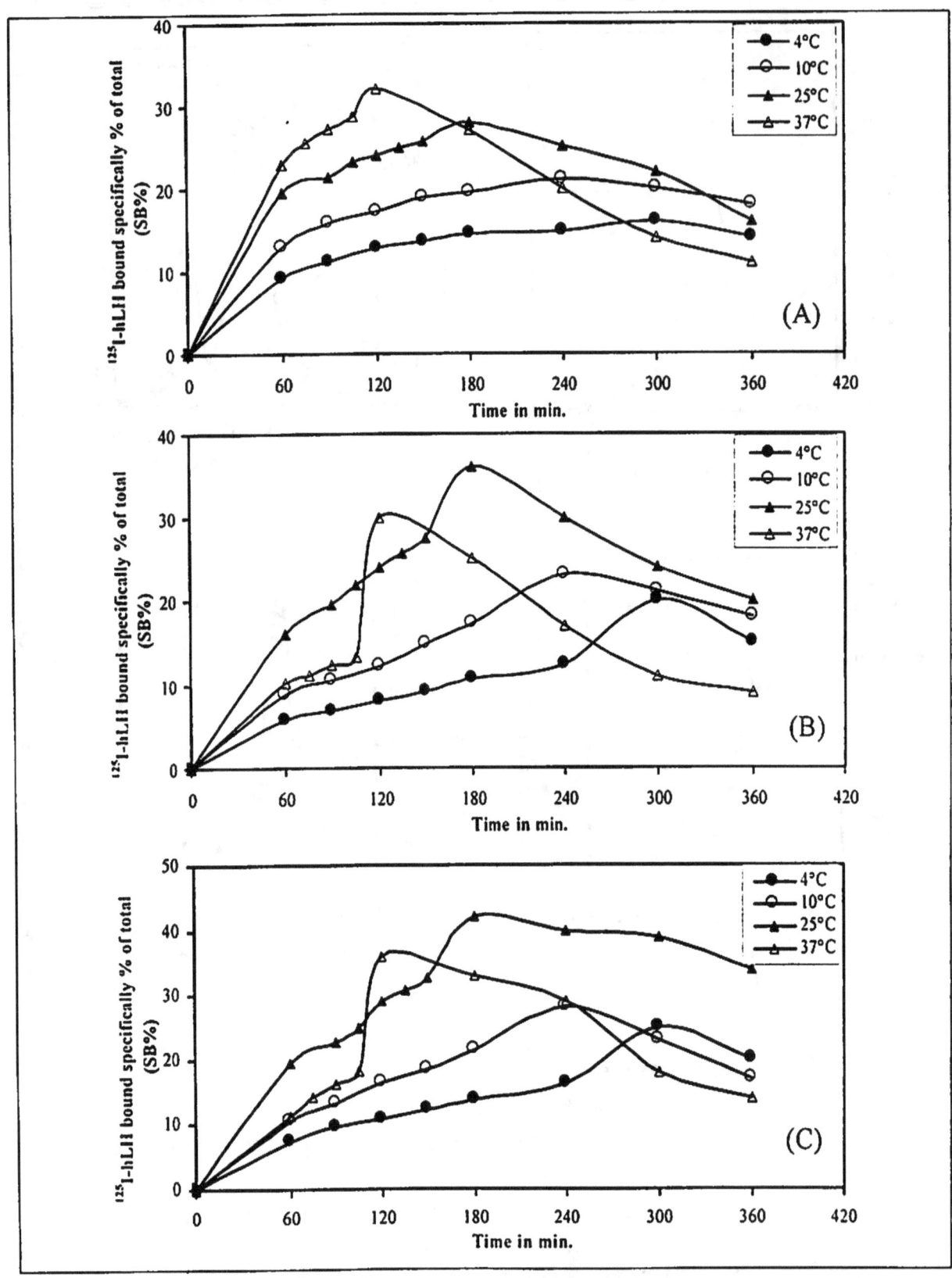

Figure (3-12): Time-course of ^{125}**I-hLH binding with its receptors in:**

A- Benign premenopausal ovarian tissue homogenate
B- Benign postmenopausal ovarian tissue homogenate
C- Postmenopausal ovarian cancer tissue homogenate

Details are described in section (2.2.7.1).

3.3.1.1 Determination of the Concentration and Affinity Constants of Luteinizing Hormone Receptors

The concentration of LH receptors and the affinity constant of the binding have been measured in pre-and postmenopausal patients with benign and malignant ovarian tumors. The experiment was carried out at the optimal conditions, which were obtained in previous experiments and was repeated at different temperatures (4, 10, 25 and 37°C). Scatchard plot analysis gave a straight line as shown in figure (3-13 A,B & C) at each temperature indicating the presence of only a single class of receptor site with specific high affinity, low capacity receptor for LH.

These specific binding sites for LH are similar to those reported previously in human ovarian follicles and corpora luteal tissue [182, 191, 202-204].

The results are summarized in table (3-2). Luteinizing hormone receptors binding capacities of premenopausal patients with benign ovarian tumors in the optimal conditions were 403.75 fmol/mg protein while of postmenopausal with benign tumor were 516.66 fmol/mg protein and of postmenopausal ovarian cancer patients were 808.33 fmol/mg protein.

A variety of normal human ovarian tissues as well as benign and malignant human ovarian neoplasms have been examined in vitro for the presence of gonadotrophin binding sites in order to determine whether gonadotrophic hormones have an effect on ovarian tumors. Epidemiologic, clinical, and animal experimental data suggested an effect of gonadotrophic hormones in the development of ovarian cancer[205]. *Sandra et. al.,* reported that epithelial tumors showing serous and mucinous differentiation demonstrated sporadic FSH as well as LH binding in both benign and malignant lesions, so gonadotropins may have been of pathogenetic significance in inducing and maintaining tumors[188]. *Rajaniemi and co-workers* found that 27% of the malignant epithelial ovarian tumors they studied were LH-receptor positive and these observations are consistent with the view that various types of ovarian cancer may be target tissues for gonadotrophins [206]. *Richard et. al.*, reported that the low level of LH binding to tumors of epithelial origin did not represent gonadotrophin-receptor binding. Thus, the common ovarian tumors of epithelial origin may not respond directly to gonadotrophin [186]. *Vinho et. al.*, reported that LH receptors were present in 68% of the menopausal patients with ovarian tumors and a negative correlation was found between serum LH levels and ovarian LH receptors [207]. *Feng et. al.*, suggested that LH stimulated the proliferation of epithelial ovarian cancer cell in about 16%, and it might play an important role in the development of epithelial ovarian cancer[208].

Immunohistochemistry confirmed the localization of LH receptor protein in the malignant epithelial tumors cells which correlated with mRNA expression[209]. In another reported survey, *Cui et. al.,* found that the level of LH receptor protein expression in patients with stage I and II of epithelial ovarian cancer was higher than that in patients with stage III and IV, but it was not significant ($p > 0.05$) and LH receptor expression did not correlate with age, lymphnode metastases, the size of residual tumor and CA-125[210].

An understanding of the significance of the LH-receptors binding sites may have practical applications in the diagnosis, treatment and prognosis of ovarian neoplasms.

Table (3-2): Concentration and affinity constant of LH receptors in three groups of serous ovarian tumor patients. Details are described in section (2.2.7.2)

Group	No. of cases	Age (year)	Binding capacity fmol/mg	ka $M^{-1} \times 10^9$
Premenopausal patients of benign ovarian tumor	4	42.0	403.75	2.29
Postmenopausal patients of benign ovarian tumor	8	56.25±2.06	516.66	2.61
Postmenopausal patients of ovarian cancer	4	57.16±1.456	808.33	3.0

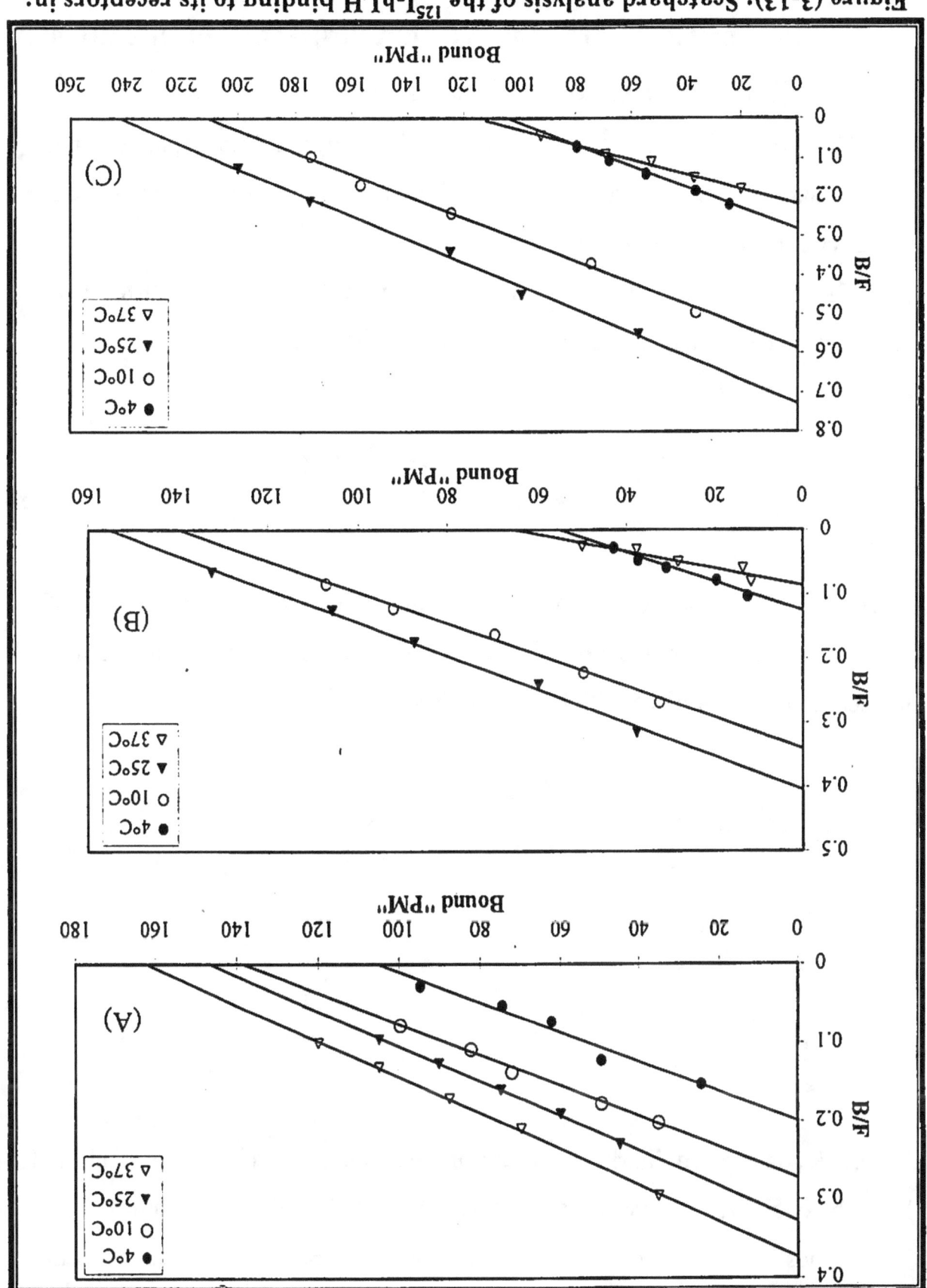

Figure (3-13): Scatchard analysis of the ^{125}I-hLH binding to its receptors in:
A: benign premenopausal ovarian tissue homogenate.
B: benign postmenopausal ovarian tissue homogenate.
C: postmenopausal ovarian cancer tissue homogenate.
Details are described in section (2.2.7.2).

3.3.1.2 Determination of Kinetic Parameters of ^{125}I-hLH Binding to Its Receptors in Pre-and Postmenopausal Patients With Serous Ovarian Tumors

The time course of ^{125}I-hLH binding to its benign and malignant receptors in pre- and postmenopausal serous ovarian tumors was carried out to describe the kinetic parameters of the binding.

The simplest proposed model representing the interaction of ^{125}I-hLH with its receptors can be expressed by the following equation:

$$^{125}\text{I-hLH} + \underset{\text{Receptor}}{\text{R}} \underset{K_{-1}}{\overset{K_{+1}}{\rightleftharpoons}} {}^{125}\text{I-hLH-R}$$

where K_{+1} is the rate of the association of ^{125}I-hLH with its receptor and K_{-1} represents the rate of the reverse reaction of the dissociation of the complex formed under the same conditions. At equilibrium:

$$K_a = \frac{[^{125}\text{I}-\text{hLH}-\text{R}]}{[^{125}\text{I}-\text{hLH}][\text{R}]} \tag{1}$$

$$K_d = \frac{[^{125}\text{I}-\text{hLH}][\text{R}]}{[^{125}\text{I}-\text{hLH}-\text{R}]} \tag{2}$$

$$\text{Thus; } K_a = \frac{1}{K_d} = \frac{K_{+1}}{K_{-1}} \tag{3}$$

Where K_a is the equilibrium constant of the association (affinity constant) and K_d is the equilibrium constant of the dissociation of (^{125}I-hLH-R) complex.

The values of K_a and maximal binding capacity (B_{max}) were calculated from scatchard plot at four different temperatures in figure (3-13 A, B and C) and table (3-3).

Results in table (3-3) show that K_a value at 25°C of postmenopausal patients with malignant tumors is about 1.1 times that of K_a value at 4°C and the K_a value of postmenopausal patients with benign tumors at 25°C is about 1.125 times of K_a value at 4°C and the K_a value at 37°C of premenopausal patients with benign tumors is about 1.2 times that of K_a value at 4°C.

The value of K_d calculated by using equation (3) shows that the lowest K_d value of (^{125}I-hLH-receptor) complex occurs at 37°C for benign premenopausal ovarian tumor homogenate and 25°C for benign and malignant postmenopausal ovarian tumors homogenate. The K_d value for the binding of LH obtained for these membranes tumors preparation is in good agreement with reports by others who used homogenates of normal rat ovaries [211], or membrane preparation of bovine corpus luteum[212]. Also it was found that in temperatures from 4°C to 25°C, the affinity of LH to its malignant receptor in postmenopausal ovarian tumor homogenate was greater than that of benign receptors in pre-and post-menopausal ovarian tumors homogenates while at 37°C, the affinity of LH to its benign receptors in premenopausal ovarian tumor homogenate was greater than that of benign and malignant receptors in postmenopausal ovarian tumors homogenates. In previous study *Lee et. al.,* reported that the binding of hLH was of high affinity with its receptors ($K_a= 0.91 \times 10^9 M^{-1}$) and the binding capacity of luteal tissue obtained from the corpus albicans of postmenopausal human ovary had a lowest value (4 fmol/mg protein) while the binding capacities of the function corpora Lutea of premenopausal human ovary varied from (14-92 fmol/mg protein)[191].

Table (3-3): The kinetic parameters of ^{125}I-hLH binding to its receptors in benign and malignant ovarian tumors. All details are described in section (2.2.7.2)

Groups	Kinetic parameters	Temperature (°C)			
		4	10	25	37
Premenopausal patients with benign ovarian tumor	Binding capacity (fmol/mg protein)	267.50	345.00	366.25	403.75
	$K_a = K_{+1}/K_{-1} \times 10^9 (M^{-1})$	1.90	2.02	2.20	2.29
	$K_d = K_{-1}/K_{+1} \times 10^{-10} (M)$	5.263	4.950	4.545	4.367
Postmenopausal patients with benign ovarian tumors	Binding capacity (fmol/mg protein)	183.33	465.00	516.66	218.33
	$K_a = K_{+1}/K_{-1} \times 10^9 (M^{-1})$	2.32	2.40	2.61	1.30
	$K_d = K_{-1}/K_{+1} \times 10^{-10} (M)$	4.348	4.166	3.831	7.692
Postmenopausal patients with malignant ovarian tumors	Binding capacity (fmol/mg protein)	355.0	710.0	808.33	383.33
	$K_a = K_{+1}/K_{-1} \times 10^9 (M^{-1})$	2.74	2.82	3.0	1.88
	$K_d = K_{-1}/K_{+1} \times 10^{-10} (M)$	3.703	3.546	3.333	5.319

Kananen et. al., reported that the equilibrium association constant ($K_a = 0.78 \times 10^9 M^{-1}$) possessed high affinity LH receptor in ovarian tumors [213], and another reported survey found that the maximal number of binding sites for LH in a purified plasma membrane preparation from rat ovary was 620 fmol/mg protein and the dissociation constant (K_d) was 5×10^{-10} M.

The Kinetic association rate constant, K_{+1}, can be determined from the time course of association of ^{125}I-hLH with its receptors and verified the order of the reaction at four different temperatures. Time-course data obtained from figure (3-12 A, B & C) can be used to confirm that the binding reaction of LH with its receptors in benign and malignant ovarian tumors homogenates following a first order kinetic reactions but due to the bimolecularity of this reaction, the following equation [214].

$$\ln(HR)_e \left[\frac{(H)_T - (HR)_t (HR)_e / (R)_T}{(H)_T - [(HR)_e - (HR)_t]} \right] = K_{+1}t \left[\frac{(H)_T (R)_T - (HR)_e}{(HR)_e} \right] \qquad (4)$$

Equation (4) can be simplified to equation (5) when the most LH remained free and only a small fraction of $(H)_T$ is bound even at equilibrium (pseudo- first order conditions)[215].

$$\ln \frac{(HR)_e}{(HR)_e - (HR)_t} = K_{+1} t \left[(H)_T (R)_T / (HR)_e \right] \qquad (5)$$

where K_{+1} is the kinetic association constant in $M^{-1} min^{-1}$; $(H)_T$ is the total molar concentration of $^{125}I\text{-hLH}$; $(R)_T$ is the total molar concentration of the hormone receptors; $(HR)_e$ is the concentration of $(^{125}I\text{-hLH-receptor})$ complex formed after time (t).

Figure (3-14 A, B & C) shows that the plotting of $\ln \dfrac{(HR)_e}{(HR)_e - (HR)_t}$ against time (t) gives a straight line with a slop equal to the observed value of first-order rate constant $(K_{obs.})$ in min^{-1}, and the association rate constant K_{+1} was calculated from the following formula:

$$K_{obs.} = K_{+1} \frac{(H)_T (R)_T}{(HR)_e} \qquad (6)$$

The half-life time of association $(t_{1/2})_{ass.}$, which represents the time needed for the formation of half amounts of the complex at equilibrium, was determined from the concentration of the complex at equilibrium and the time course curve, while the half-life time of dissociation $(t_{1/2})_{diss.}$ was determined from:

$$(t_{1/2})_{diss.} = \frac{\ln 2}{K_{-1}} = \frac{0.693}{K_{-1}}$$

The K_a values were also obtained from equation (3). Figure (3-14 A, B& C) represents the kinetics of complex formation between $^{125}I\text{-hLH}$ and its receptors in two groups of benign ovarian tumor homogenates of pre and postmenopausal patients and one group of malignant ovarian tumor homogenate of postmenopausal patients at different temperatures.

The results revealed that the association rate constant K_{+1} at 37°C was higher than that of 4°C by approximately 2.12 folds for benign premenopausal tumors and K_{+1} at 25°C was higher than that of 4°C by approximately 1.4 folds for benign postmenopausal tumors while, the K_{+1} value at 25°C was higher than that of 4°C by approximately 1.5 folds for malignant postmenopausal tumors and also it was found that K_{+1} values at 4, 10, 25 and 37°C for the two groups of benign tumors were higher than that of malignant tumors at each temperature as shown in table (3-4).

The slow rate dissociation in malignant group may insure the hypothesis that ovarian cancer may lead to a prolongation of the half-life of LH[216]. This prolongation in the half-life can lead to regulate events leading to ovarian cancer cell proliferation and thereby, ovarian cancer cell under LH control can synthesize and secrete their own growth factors that can autostimulate the ovarian cancer cell[58]. The high rate dissociation in benign groups can be viewed as an indirect support to the concept that LH is involved in the proliferation of human benign ovarian tumors[215]. Thus, the results revealed that LH had a role in the growth of ovarian tumor cell in malignant group more than that in the benign groups.

The values of K_{-1} were obtained also from the values of K_a which have been estimated at the four different temperatures investigated. The K_{-1} was determined from the equation (3). Table (3-4) shows that K_{-1} increases with the elevation of temperature. Thus, when the reaction temperature was increased from 4°C to 37°C, the value of the dissociation constant increased approximately 1.76 folds for benign premenopausal tumors receptors and 2.4 folds for benign postmenopausal tumors receptors while that of malignant postmenopausal tumors receptors the K_{-1} increased approximately 2.1 folds.

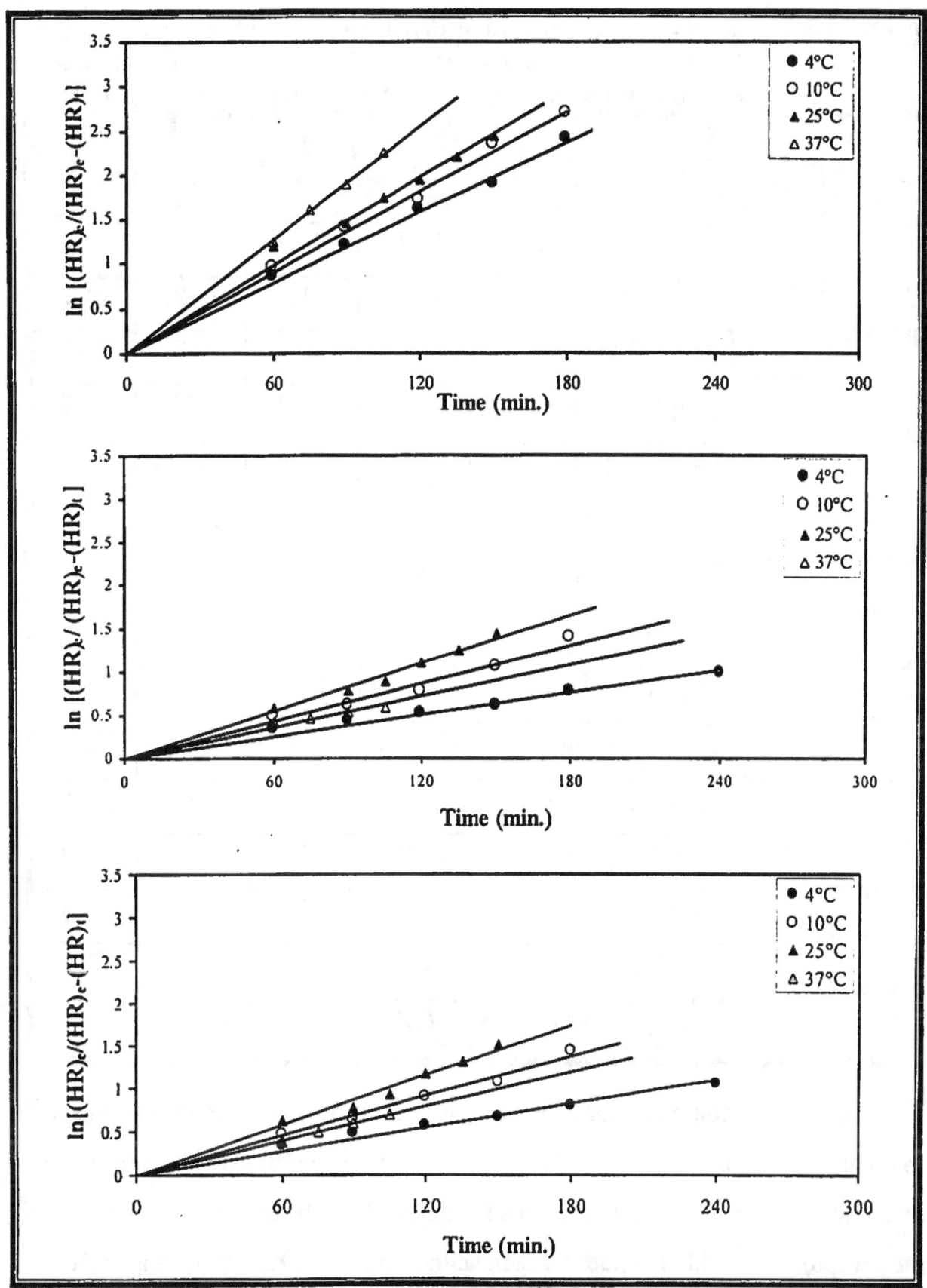

Figure(3-14):Pseudo-first order kinetics of ^{125}**I-hLH binding with its receptors in**
      A: benign premenopausal ovarian tissue homogenate,
      B: benign postmenopausal ovarian tissue homogenate,
      C:postmenopausal ovarian cancer tissue homogenates, at four different temperature.
Details are described in section (2.2.7.3)

Table (3-4): The effect of different temperature on the kinetic parameters of LH binding to its receptors in benign and malignant ovarian tumors. All details are described in section(2.2.7.3)

Groups	Kinetic parameters	Temperature (°C)			
		4	10	25	37
Premenopausal patients with benign ovarian tumors	$K_{+1} \times 10^{7} (M.min)^{-1}$	1.973	2.29	3.14	4.18
	$K_{-1} \times 10^{-2} (min^{-1})$	1.038	1.134	1.427	1.825
	$K_{a} \times 10^{9} (M^{-1})$	1.90	2.02	2.20	2.29
	$(t_{1/2})_{ass.}$ (min)	51.0	45.0	39.0	36.0
	$(t_{1/2})_{diss.}$ (min)	66.763	61.111	48.563	37.972
Postmenopausal patients with benign ovarian tumors	$K_{+1} \times 10^{7} (M.min)^{-1}$	1.527	1.64	2.136	2.089
	$K_{-1} \times 10^{-2} (min^{-1})$	0.6581	0.6833	0.8183	1.6069
	$K_{a} \times 10^{9} (M^{-1})$	2.32	2.40	2.61	1.30
	$(t_{1/2})_{ass.}$ (min)	150.0	96.0	75.0	105.0
	$(t_{1/2})_{diss.}$ (min)	105.303	101.420	84.687	43.126
Postmenopausal patients with malignant ovarian tumors	$K_{+1} \times 10^{7} (M.min)^{-1}$	1.11	1.285	1.671	1.607
	$K_{-1} \times 10^{-2} (min^{-1})$	0.405	0.455	0.557	0.854
	$K_{a} \times 10^{9} (M^{-1})$	2.74	2.82	3.0	1.88
	$(t_{1/2})_{ass.}$ (min)	150.0	90.0	72.0	90.0
	$(t_{1/2})_{diss.}$ (min)	171.11	152.30	124.42	81.15

The association rate constants for the interaction of hLH to its benign and malignant receptors of postmenopausal ovarian tumors homogenates determined at 25°C compare favorably with those obtained from the interaction of ovarian rat receptors with hLH under the same conditions (The K_{+1} value ranged from 15.0 to 22.8 x 10^{7} $M^{-1}.min^{-1}$).[189] This numerical difference may be attributed to the different types of receptor sources used and to differences in the structure of these receptors.

3.3.2 The Thermodynamic of the Binding of ^{125}I-hLH to Its Receptors in Benign and Malignant Ovarian Tumors

3.3.2.1 Thermodynamic Parameters of Standard State

Figure (3-15) represents the dependence of the equilibrium binding constant (i.e., affinity constant) for the binding of ^{125}I-hLH to its receptors in pre-and postmenopausal ovarian tumor homogenate on the temperature (*Van't Hoff plot*).

The results indicated that $\Delta H°$ in general had small values and nearly close to zero, their positive sign ascertain that the reaction was nearly endothermic. The $\Delta H°$ value in the case of benign premenopausal ovarian tumor receptors was higher than that in the cases of benign and malignant postmenopausal ovarian tumors receptors, so more energy was needed in case of benign premenopausal ovarian tumor receptors for the reaction (binding) to occur.

The negative values of $\Delta G°$ reflect the stability of the complex, hence the high affinity of the reactants. The high negative values of $\Delta G°$ for the binding reactions are controlled by high positive $\Delta S°$ values as shown in table (3-5). So, our system is characterized by the sole contribution of $\Delta S°$ to the stability of the complexes formed, while $\Delta H°$ has little or no effect [217].

The high value of positive $\Delta S°$ suggests that the reaction spontaneity was entropically driven. Entropy was the driving force for the occurrence of the binding reaction. This indicates that the hydrophobic interactions played an important role in stabilizing the complex[218].

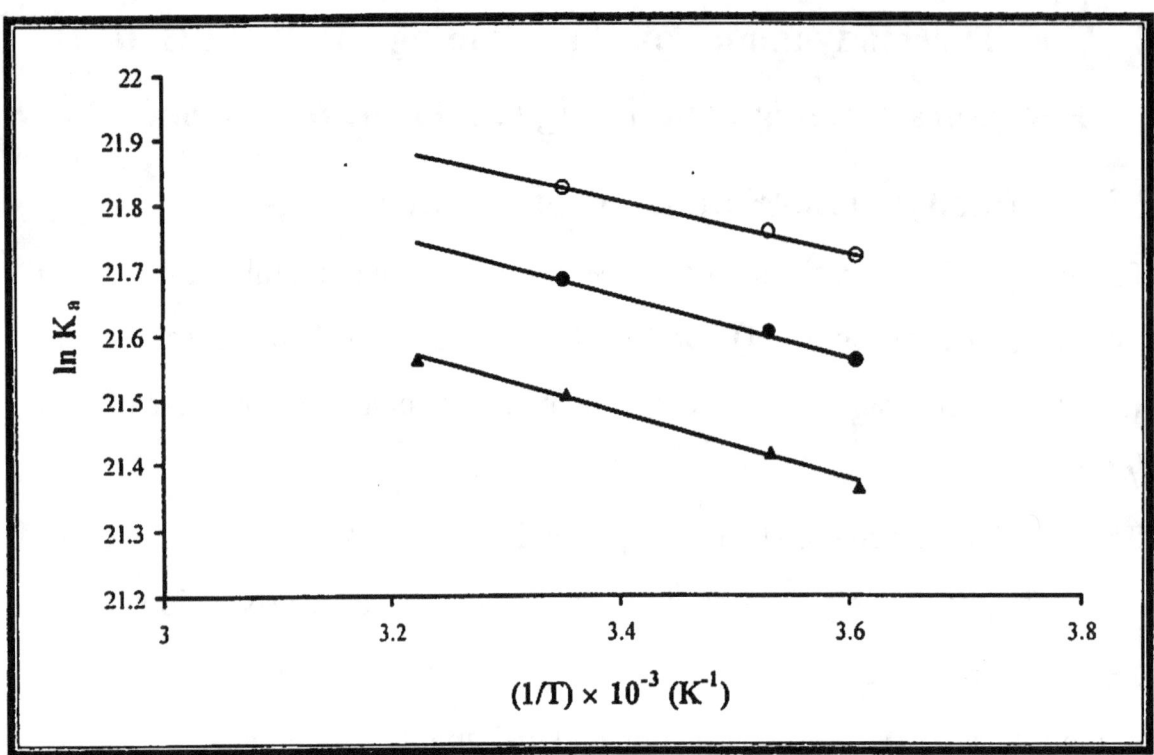

Figure (3-15): Van't Hoff for the ^{125}I-hLH binding to its receptors in
  ▲　benign premenopausal ovarian tissue homogenate
  ○　benign postmenopausal ovarian tissue homogenate
  ●　postmenopausal ovarian cancer tissue homogenate
Details are described in section (2.2.7.4)

Table (3-5): Thermodynamic parameters at standard state of LH binding to its receptors in pre-and postmenopausal ovarian tumor homogenate. All details are described in section (2. 2.7.4).

Group	Thermodynamic parameters	Temperature (°C)			
		4	10	25	37
Premenopausal patients with benign ovarian tumors	$\Delta H°$(KJ/mole)	+4.204	+4.204	+4.204	+4.204
	$\Delta G°$(KJ/mole)	-49.226	-50.436	-53.320	-55.569
	$\Delta S°$(J/mole.K)	+192.7976	+192.9855	+192.9493	+192.7353
Postmenopausal patients with benign ovarian tumors	$\Delta H°$(KJ/mole)	+3.966	+3.966	+3.966	+3.966
	$\Delta G°$(KJ/mole)	-49.686	-50.842	-53.743	-54.109
	$\Delta S°$(J/mole.K)	+193.5986	+193.5789	+193.5699	+187.260
Postmenopausal patients with malignant ovarian tumors	$\Delta H°$(KJ/mole)	+3.407	+3.407	+3.407	+3.407
	$\Delta G°$(KJ/mole)	-50.070	-51.222	-54.088	-55.060
	$\Delta S°$(J/mole.K)	+192.967	+192.947	+192.852	+188.524

The small positive $\Delta H°$ may indicate a favorable interactions between groups within both LH and its receptors. These include the non-covalent interactions which are fundamentally electrostatic in nature such as charge-charge interactions which occurs in both LH and its receptors in ovarian tumor homogenate, other types of interactions include charge-dipole, dipole-dipole, charge-induced dipole, dipole-induced dipole and hydrogen bond. The sum of these types of interactions can yield some stabilization to the folded structure of the complex. So, the negative value of $\Delta G°$ showed that the overall reaction was energetically favorable in the direction of complex formation.

Calculation of the thermodynamic property, $\Delta G°$, for LH binding to its receptors is in good agreement with the value reported for LH – ovarian rat receptors interaction ($\Delta G°$=-14.0 Kcal/mol= -58.578 KJ/mol at 25°C)[189]

3.3.2.2. Thermodynamic Parameters of Transition State

According to the transition state theory, the interaction of two proteins leads to the formation of an activated complex (transition state), then the formation of the final product:

$$^{125}I - hLH + R \rightarrow [^{125}I - hLH - R] \rightarrow {}^{125}I - hLH - R$$

an activated complex

(transition state)

The transition state thermodynamic parameters ΔH^*, ΔG^*, ΔS^* and Ea could be determined from Arrhenius equation. Figure (3-16) shows the dependence of the association rate for the binding of ^{125}I-hLH to its receptors in ovarian tumor homogenate on temperature (*Arrhenius plot*).

The high positive value of ΔG^* indicated that the formation of an activated [hLH-R] complex was a non spontaneous process and required a lot of energy (equal to Ea) to overcome the transition state energy barrier and giving the final product, whereas the high negative ΔS^{*} revealed that the activated complex had a more ordered structure than the reactant species ($\Delta S^{*} < 0$) as shown in table (3-6). The positive values of ΔG^{*} is mainly attributed to the decrease in entropy of the transition state ($\Delta S^{*} < 0$). In addition, the positive value of ΔH^{*} shows that the heat content of the activated complex is more than that of isolated species[219].

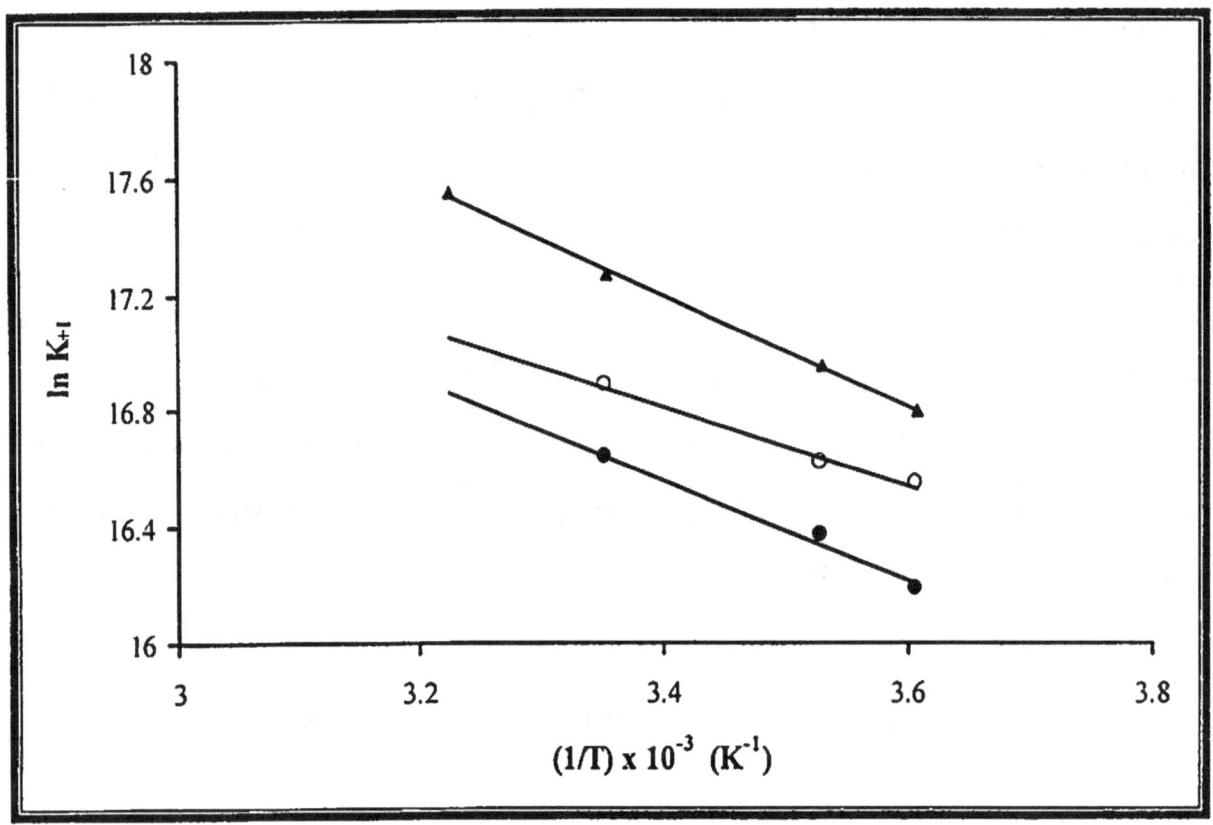

Figure(3-16): Arrhenius plot for the ^{125}I-hLH binding to its receptors in:
    ▲   benign premenopausal ovarian tissue homogenate
    ○   benign postmenopausal ovarian tissue homogenate
    ●   postmenopausal ovarian cancer tissue homogenate
Details are described in section (2.2.7.4)

Table (3-6): Thermodynamic parameters at transition state of LH binding to its receptors in benign and malignant ovarian tumors. All details are described in section (2.2.7.4)

Group	Thermodynamic parameters	Temperature (°C)			
		4	10	25	37
Premenopausal patients with benign ovarian tumors	Ea (KJ/mole)	16.00	16.00	16.00	16.00
	ΔH^* (KJ/mole)	13.696	13.646	13.521	13.42
	ΔG^* (KJ/mole)	29.00	29.329	30.228	30.810
	ΔS^* (J/mole. K)	-55.223	-55.392	-56.039	-56.073
Postmenopausal patients with benign ovarian tumors	Ea (KJ/mole)	11.226	11.226	11.226	11.226
	ΔH^* (KJ/mole)	8.922	8.872	8.747	8.647
	ΔG^* (KJ/mole)	29.592	30.115	31.184	32.598
	ΔS^* (J/mole.K)	-74.586	-75.029	-75.259	-77.229
Postmenopausal patients with malignant ovarian tumors	Ea (KJ/mole)	14.178	14.178	14.178	14.178
	ΔH^* (KJ/mole)	11.874	11.824	11.699	11.599
	ΔG^* (KJ/mole)	30.327	30.689	31.792	33.274
	ΔS^* (J/mole. K)	-66.586	-66.630	-67.397	-69.890

Determination of thermodynamic parameters of the binding reaction using equilibrium data gives an overall idea about the nature of forces controlling complex formation. Comparison of the values of transition state with those of standard state in tables (3-5) and (3-6) lead us to choose a thermodynamic model shown in figure (3-17). Our model proposes that the formation of the (^{125}I-hLH-receptor) complex undergoes three thermodynamic states. Thermodynamic state (A) represents the initial energy level of the isolated ^{125}I-hLH and its receptor(R). In thermodynamic state (B), the two components have come together and mutually penetrated their hydration sphere to form a partially immobilized hydrophobically associated species. Thermodynamic state (C) represents the fully interacting complex (^{125}I-hLH-R). In step 1 of the reaction, the binding of ^{125}I-hLH to its receptor was associated with positive ΔG^* value. This indicates that the initial step of the reaction requires input of energy for the system. The negative entropy change ΔS^* for this step of the reaction reflects the change of the ^{125}I-hLH-R transition complex to a more ordered structure. The positive $\Delta H^{\bullet}$ value shows that the heat content of the activated complex is more than that of the isolated species. Partial immobilization of the hydrophobically associated complex formed, in step 1, occurs when isolated hydrated species ^{125}I-hLH and receptor (R) interact partially so that there is a mutual penetration of their hydration layers to form the activated complex. This hydrophobic association is a result of the tendency of water to form a more ordered structure in the vicinity of non-polar hydrocarbon groups (e.g, the side chains of the amino acids phenyl alanine, Leucine and tryptophan) this means that hydrophobic amino acid side chains which were previously accessible to solvent in the isolated species became buried upon complex formation. In step 2, the activated complex participates in further interactions, giving the fully interacting complex (^{125}I-hLH-R). It is proposed that the formation of a LH-receptor complex occurs in two steps, the first step stabilized the complex by hydrophobic interactions and the second step stabilized it by short range interactions such as electrostatic interactions, hydrogen bonding and van der Waals' interactions[220].

Hydrophobic interactions contribute to the stability of the complex via large decrease in the excited entropy change ($\Delta S^{\bullet} < 0$), while electrostatic interactions, hydrogen bonding and Van der Waals' interactions stabilize the complex via high increase in the standard entropy change ($\Delta S^{o} > 0$). [220,221]

The thermodynamic data from this study indicate that the binding of ^{125}I-hLH to its receptors are entropically driven and come in agreement with the concept that hydrophobic and short-range interactions have an important role in ^{125}I-hLH-R interactions.

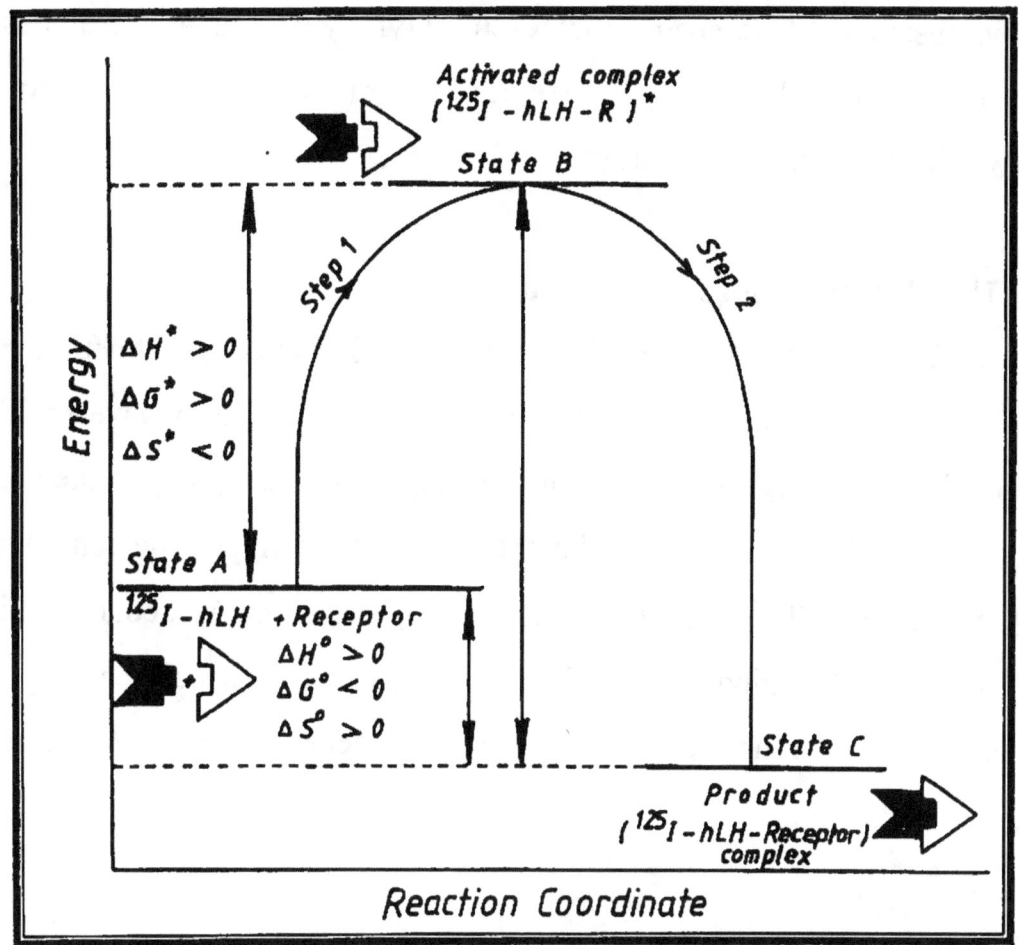

Figure (3-17): General energy diagram and thermodynamic model applied to the interaction of ^{125}I-hLH with its receptor in ovarian tumors

3.4 Spectroscopic Studies on hLH

3.4.1 The U.V Spectrum of hLH, hLH Antibody, (hLH-Antibody) Complex and (^{125}I-hLH-Receptor) Complex

- **The U.V Spectrum of hLH**

Figure (3-18A) illustrates the U.V spectrum of hLH (provided by hLH RIA kit, DPC-USA) at neutral pH. The spectrum shows that the λmax for hLH is consisted of two maximum wavelength, λmax_1, at 278.4 nm, and λmax_2, at 228.6 nm. The absorption at 278.4 nm is characteristic of the side chain chromophore of tyrosyl residues, while the absorption at 228.6 nm is due to the amide group in the polypeptide bond of the hLH molecule with contribution of the histidyl residues[222]. It seems that each of tyrosyl and histidyl residues in hLH molcule is located in away, that part of it is on the surface of the protein molecule while the other part is buried.

- ## The U.V Spectrum of hLH Antibody

Figure (3-18B) illustrates the U.V spectrum of hLH antibody at neutral pH. The U.V spectrum of hLH antibody consists of one maximum wavelength at 204.6 nm which is assigned to the amide group only in the polypeptide bond, and a shoulder at 233.4 nm which is characteristic to the polypeptide bond in the antibody molecule with contribution of the histidyl residues. At a concentration of (1.3 g/L) of hLH antibody and 0.5 cm cell thickness the specific absorption coefficient $as_{204.6}$ was found to be 2.623 $g^{-1}.cm^{-1}.L$ according to *Lambert Beer's law* [172].

- ## The U.V Spectrum of (hLH-Antibody) Complex

Figure (3-18C) shows the U.V spectrum of (hLH-antibody) complex at neutral pH. The spectrum shows that there is one maximum wavelength at 207.0nm compared to the U.V spectrum of hLH and the antibody, Figure (3-18 A & B). The λmax (278.4 and 228.6 nm) for the hLH molecule and λmax (204.6 and 233.4nm) for the antibody molecule were disappeared. The absorption at 207.0 nm is a characteristic of the amide group in the polypeptide bond of the complex molecule [223]. The disappearance of the absorption at 278.4 nm of the tyrosyl residues indicates that the tyrosine is in the active site of hLH molecule [224].

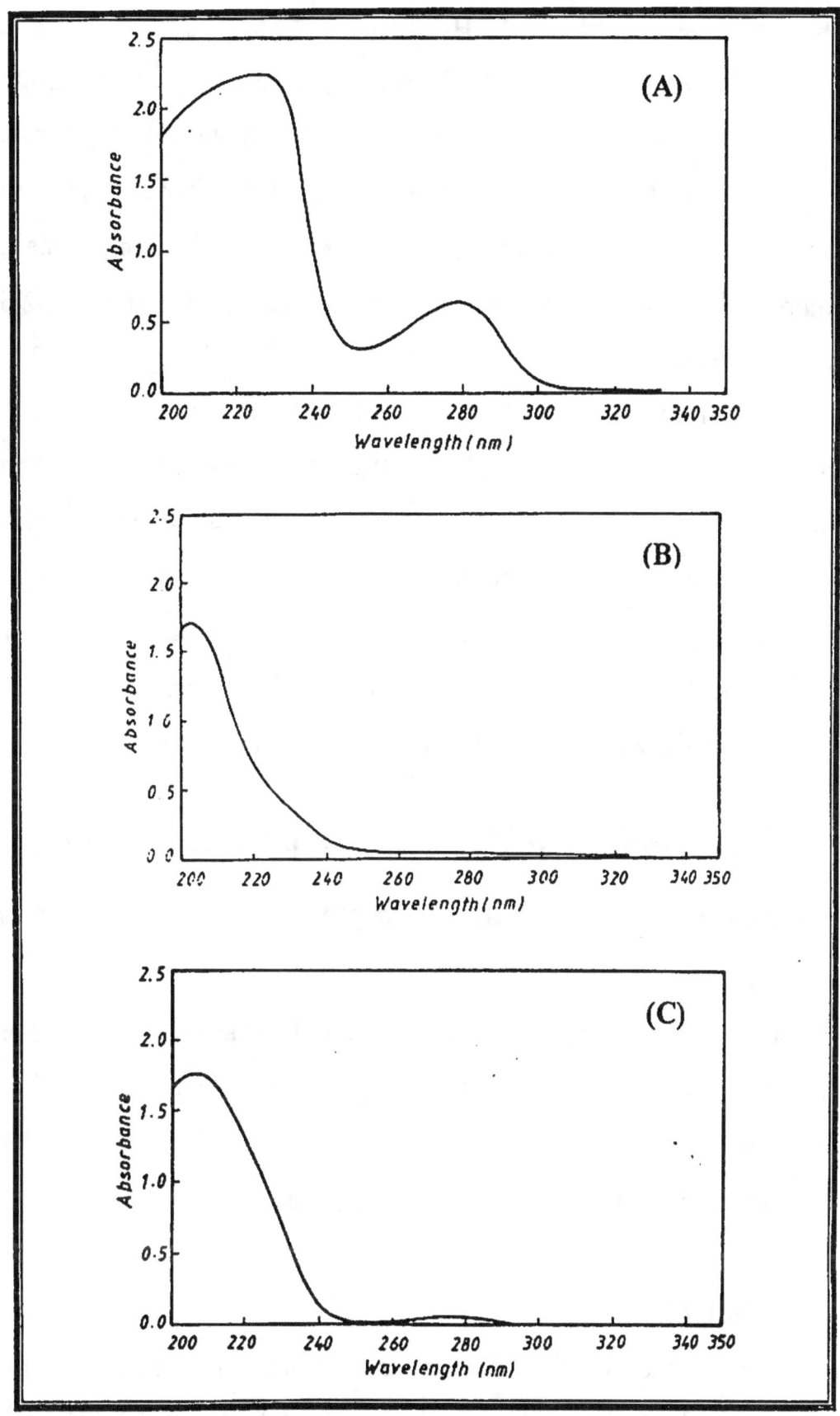

Figure (3-18): The U.V spectrum of
A-hLH,   B-hLH antibody,
C- (hLH-antibody) complex, at neutral pH
Details are described in section (2.3.1.1).

- ### The U.V Spectrum of (^{125}I-hLH-Receptor) Complex

The U.V spectrum of (^{125}I-hLH-receptor) complex in both benign and malignant postmenopausal serous ovarian tumors homogenates were measured at pH (8.2). The spectrum shows two maximum wavelength, λmax_1 at 227.6 and 226.2nm and the other λmax_2 at 192.6 and 194.0 nm for benign and malignant postmenopausal tumors respectively. The wavelength at λmax_1 192.6 and 194.0nm is assigned to the tyrosyl residues [224], and the absorption at 227.6 and 226.2nm arise from electronic transitions in the polypeptide backbone with contribution of the histidyl residues [222], while the U.V spectrum of the complex for benign premenopausal ovarian tumor homogenate was measured at pH (7.4). The spectrum shows two maximum wavelength, λmax_1 at 276.0 nm and the other λmax_2 at 224.4 nm. The first peak λmax_1 is assigned to the tyrosyl residues [224], and the absorption at 224.4 nm arise from electronic transitions in the polypeptide bond with contribution of the histidyl residues.

3.4.2 Factors Affecting the Absorption Properties of hLH, hLH Antibody,(hLH-Antibody)Complex and(^{125}I-hLH-Receptor) Complex

The absorption spectrum of a chromophore is determined by the chemical structure of the molecule. However, a large number of environmental factors produce detectable change in λmax and absorbance. Environmental factors consist of pH, the polarity of the solvent or neighboring molecules [224].

3.4.2.1 pH Effect

The pH of the solvent determines the ionization state of ionizable chromophores [224]. Table (3-7A) shows the λmax values of hLH, hLH antibody and (hLH-antibody) complex at different pH (3.8, 7.3 and 12.6). Figure (3-19 A, B & C) shows these spectrum.

From Table (3-7A), it seems that hLH in acidic region at pH 3.8 there is a slight shift to a shorter λmax 278.0, 222.4 nm for the tyrosyl residues and the polypeptide bond respectively. This decrease was associated with the decrease of the absorbancy. The U.V spectrum of hLH antibody and (hLH-antibody) complex in acidic region (pH 3.8) shows a slight shift to a shorter λmax 203.0, 204.6 respectively, for the amide groups in the polypeptide bond only. While λmax$_1$ at 233.4 nm in hLH antibody spectrum disappeared at pH 3.8.

The spectrum shifts of proteins produced by pH cannot be simply attributed to the inductive effects at vicinal charges, such spectral changes must therefore be attributed mainly to rearrangement of secondary and tertiary structure, although the possibility of field effects due to unusually close conjunction of charges aromatic groups is not excluded [225].

Table (3-7A): The effect of pH on the λmax of U.V spectrum of hLH, hLH antibody and (hLH-antibody) complex.
Details are described in section (2.3.1.1)

pH	hLH λmax (nm)	hLH Antibody λmax (nm)	(hLH-Antibody) Complex λmax (nm)
3.8	278.0, 222.4	203.0	204.6
7.8	278.4, 228.6	233.4, 204.6	207.0
12.8	287.6, 220.6	215.8	218.8

When the pH was increased to pH (12.6) there was a significant shift to a longer wavelength (red shift) for tyrosyl residues in the U.V spectrum of hLH. The increase in λmax$_1$ 287.6 nm was due to the dissociation of the phenolic OH of the tyrosine (pKa 10.07), while λmax$_2$ 220.6 nm of the polypeptide bond with contribution of histidyle residues could be due to a conformation change in the protein [223].

Since the spectral change is a function of pH, this indicates that the tyrosine residues are on the surface of the protein (hLH) [225]. The U.V spectrum of hLH antibody and (hLH-antibody) complex in basic region at pH (12.6) shows a shift to a longer wavelength (red shift) for the amide groups in the polypeptide bond, λmax 215.8 and 218.8 nm respectively, while the λmax 233.4 nm for the hLH antibody molecule were disappeared at pH 12.6, these could be due to a conformational changes in the protein molecule.

The effect of pH on the U.V spectrum of (^{125}I-hLH-receptor) complex was studied. Table (3-7B) shows the λmax values for this complex at different pH(7.4, 8.2, and 9.0). At pH 9.0, the U.V spectrum of (^{125}I-hLH-receptor) complex in benign and malignant postmenopausal ovarian tumors homogenates shows that λmax$_2$ was increased from (192.6, 194.0 nm to 200.2 and 205.8 nm) with a decrease in the absorbance, while λmax$_1$ (227.6 and 226.2 nm) remained constant and a new λmax was obtained at 276.8 and 274.4 nm which is assigned to tyrosine. The appearance of this new λmax indicate that the protein was defolded which is due to the change in the secondary and tertiary structure of the protein that being the tyrosyl residues to expose to absorbance [224].

The U.V spectrum of (^{125}I-hLH-receptor) complex in benign premenopausal ovarian tumors at pH 9 shows that the λmax$_1$ 276.0 nm was remained constant, while the λmax$_2$ increased from 224.4 to 231.4 nm with a decrease in the absorbance and a new λmax was obtained at 200.8 nm that is assigned to the amide group in polypeptide bond.

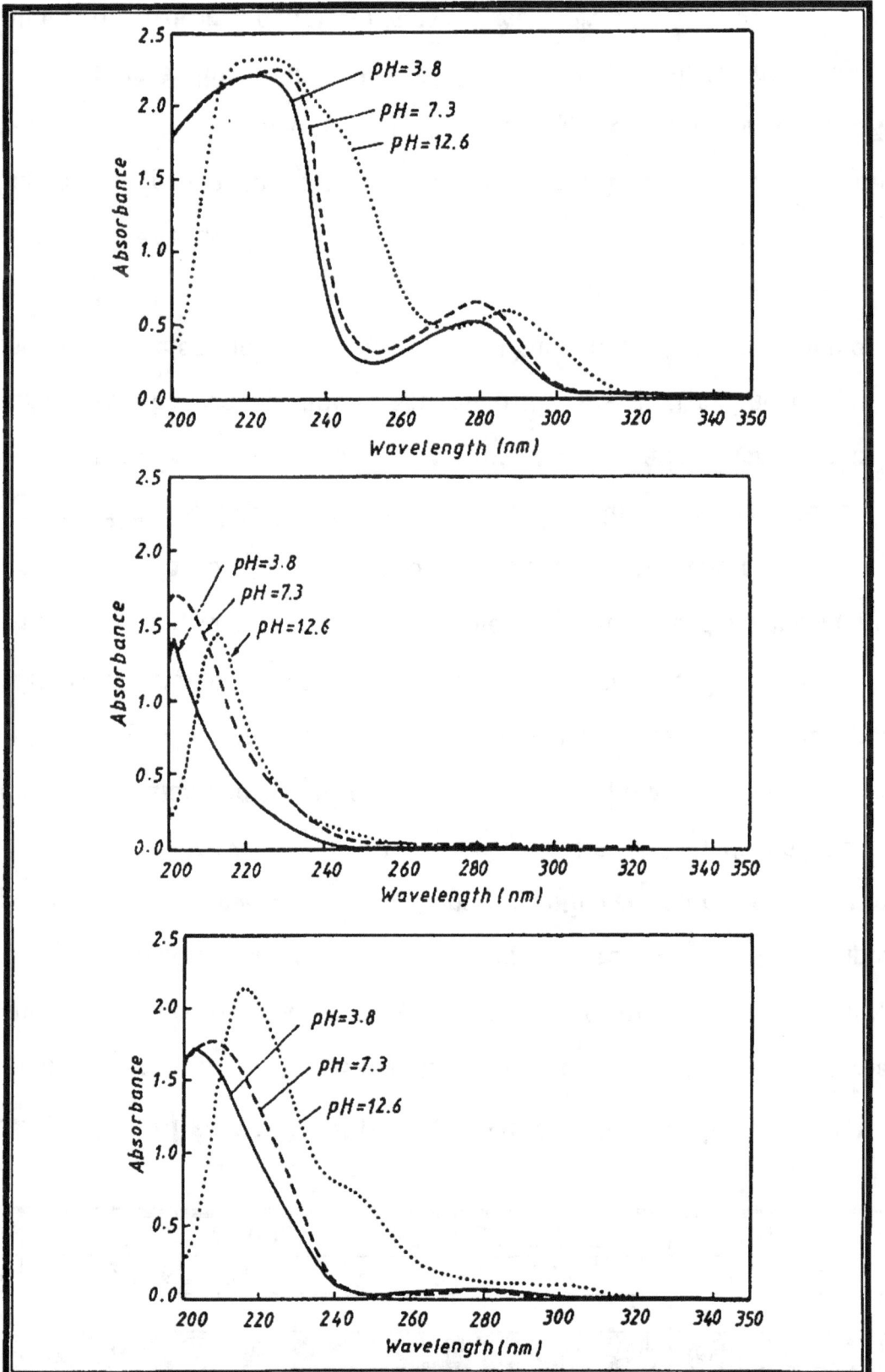

Figure (3-19): The pH effect on U.V spectrum of
A-hLH,
B-hLH antibody,
C- (hLH-antibody) complex, at pH (3.8, 7.3 and 12.6)
Details are described in section (2.3.1.1).

Table (3-7B): The effect of pH on the λmax of U.V spectrum of (^{125}I-hLH-receptor) complex in pre-and postmenopausal ovarian tumors. Details are described in section (2.3.1.1)

pH	postmenopausal patients with benign ovarian tumor. λmax (nm)	postmenopausal patients with malignant ovarian tumor. λmax (nm)	premenopausal patients with benign ovarian tumor. λmax (nm)
7.4	-	-	276.0, 224.4
8.2	227.6, 192.6	226.2, 194.0	-
9.0	276.8, 227.6, 200.2	274.4, 226.2, 205.8	276.0, 231.4, 200.8

3.4.2.2 Effect of solvent polarity (solvent perturbation)

The determination of whether an amino acid is internal or external by measuring the spectra of a protein in a polar and non-polar solvent is called the solvent perturbation method. In fact, proteins are rarely studied in completely nonpolar solvents because most proteins are either insoluble or denatured in these solvents. However, significant solvent effects can be induced by use of a mixture of water and a substance of a reduced polarity such as glycerol, polyethylene glycol, ethanol, dimethyl sulfoxide and urea [224]. Several spectral changes were obtained in the presence of these perturbants, like the alteration of the λmax position and intensities of protein spectrum, and the appearance of new chromophores on the surface of protein molecule. These chromophores were embedded in an interior region of the protein in the absence of the solvent.

One of the main assumptions of the solvent perturbation technique is that solvents alter the peak positions and intensities by altering the energy and probability of electronic transitions. Other considerations include the following: [226-228]

a- polarization effect.

b- changes in permanent dipole moment during excitation i.e., the dipole hydrogen banding, which will tend to produce either a short wave or long wave shift depending on the nature of the electronic transition and weather the solute is the hydrogen donor or hydrogen acceptor.

1- **The Effect of 20% Ethanol, Polyethylene Glycol (PEG), Glycerol and Dimethyl Solfoxide (DMSO) on hLH, hLH Antibody (hLH-Antibody) Complex and (^{125}I-hLH-Receptor) Complex U.V Spectrum.**

Table (3-8A) shows the λmax values of hLH spectrum in 20% of different solvents at neutral pH. When comparing the values of λmax in table (3-8A) with those in table (3-7A) at pH 7.3, it seems that the presence of 20% ethanol has an effect on the position of the λmax$_1$ which is specific for the tyrosyl residues in hLH molecule, so there is a significant blue shift in this λmax. This shift is due for n→π* transition which occurs at a shorter wavelength in polar hydroxylic solvent like ethanol[224]. On the other hand, there is a significant red shift in the λmax$_2$ that is specific for the amide group of the polypeptide bond with contribution of histidyl residues. These alteration in the positions at λmax$_1$ or λmax$_2$ are all due to the intermolecular hydrogen bonding between the tyrosyl residue or the amide bonds in the hLH molecule with ethanol. Hence additional bands start to appear at longer or shorter wavelength[229]. The presence of 20% PEG shows no effect on the λmax of tyrosyl residues as shown in table (3-8A), and there is a slight red shift (2 nm) in λmax$_2$. When the (20%) Glycerol was used, a slight blue shift is noticed in λmax$_1$ of tyrosyl residues and a slight red shift (6 nm) in λmax$_2$ of amide group in the polypeptide bond with contribution of histidyl residues. The presence of 20% DMSO has a great effect on the protein structure of hLH molecule because of the disappearance of λmax$_2$ 228.6nm and a new λmax were obtained at 250.2 nm which is assigned to the π→π* transition of the aromatic ring of phenylalanine[229], while there is a red shift in λmax$_1$ from 278.6 to 284.6 nm, this red shift related to tyrosine residue, i.e. the other part of tyrosine which was buried was brought to expose to absorbance in the presence of DMSO. Figure (3-20A) shows the spectrum of hLH in the presence of 20% of different solvents.

Table (3-8A):   The effect of 20% of ethanol, PEG, glycerol and DMSO on λmax of hLH U.V spectrum at neutral pH. Details are described in section (2.3.1.2).

Solvent	λmax (nm)
Ethanol	269.0, 234.6
PEG	278.6, 230.2
Glycerol	277.8, 234.6
DMSO	284.6, 250.2

Table (3-8 B) shows the λmax values of hLH-antibody spectrum in the presence of 20% ethanol, PEG, glycerol and DMSO at neutral pH. There was a significant red shift in λmax$_2$ 204.6 to 217.8 that is specific for the amide group of the polypeptide bond. On the other hand, λmax$_1$ at 233.4 nm disappeared in the presence of 20% ethanol, when PEG (20%) was used there were a blue shift in the λmax$_1$ 233.4 to 228.4 nm and a red shift in λmax$_2$ 204.6 to 209.0 nm of the amide bond. Glycerol at a concentration of 20% showed a significant red shift in λmax$_1$ from 204.6 to 220.4 nm while λmax$_2$ disappeared, this may be due to a conformational changes in the protein molecule. In the case of DMSO (20%). λmax$_1$ increased from 233.4 to 235 nm, while λmax$_2$ was disappeared. Figure (3-20 B) shows the effect of above solvents on the λmax of hLH antibody spectrum.

Table (3-8 B):   The effect of 20% ethanol, PEG, glycerol and DMSO on λmax of hLH antibody U.V spectrum at neutral pH. Details are described in section (2.3.1.2)

Solvent	λmax (nm)
Ethanol	217.8
PEG	228.4, 209.0
Glycerol	220.4
DMSO	235.0

Table (3-7C) and figure (3-20C) show the effect of different solvents on the (hLH-antibody) complex at neutral pH, (λmax =207.0 nm shown in previous experiments), was shifted towards longer wavelength in both of ethanol and PEG at a concentration of (20%). These shifts are attributed to the amide group in polypeptide bond with the contribution of histidyl residues. Glycerol (20%) showed an increase in λmax from 207.0 to 225.2 nm, and a new λmax at (279.0 nm) appeared. The new peak is related to tyrosine residue which may be affected with neighboring groups by some hydrophobic interactions. In the case of DMSO (20%), two peaks were obtained, at (285.0 and 252.6 nm). The λmax at 285.0 nm is related to the $\pi \rightarrow \pi^*$ transition of the aromatic ring of tryptophan and this attributed to a change in the protein structure that brought tryptophan to expose to absorbance in the presence of DMSO, while the λmax at 252.6 nm is assigned to phenylalanyl residue. The disappearance of λmax at 207.0 nm and the presence of two new chromophores may be due to that (hLH- antibody) complex defolded in the presence of DMSO.

Table (3-8C): **The effect of 20% ethanol, PEG, glycerol and DMSO on λmax of (hLH-antibody) complex U.V spectrum at neutral pH. Details are described in section (2.3.1.2)**

Solvent	λmax (nm)
Ethanol	223.4
PEG	220.4
Glycerol	225.2, 279.0
DMSO	252.6, 285.0

Finally, Table (3-8D) shows the effect of different solvents on the (^{125}I-hLH -receptor) complex U.V spectrum. Ethanol 20% showed that there was a slight blue shift in λmax$_1$ and λmax$_2$ at pH 8.2 in the case of benign and malignant postmenopausal ovarian tumor, while in benign premenopausal ovarian tumor, a new λmax was obtained at 199.8 and 208.8 nm at pH 7.4 which were assigned to tyrosine and the amide group in polypeptide bond respectively.

While λmax at 276.0 nm was decreased to 271.0 nm and the λmax at 218.6 nm was disappeared.

Table (3-8D): The effect of 20% ethanol, PEG, Glycerol and DMSO on the λmax of (^{125}I-hLH-receptor) complex U.V. spectrum. Details are described in section (2.3.1.2)

Solvent	Postmenopausal patients with benign ovarian tumor. λmax (nm)	Postmenopausal patients with malignant ovarian tumor. λmax (nm)	Premenopausal patients with benign tumor. λmax (nm)
ethanol	224.9, 190.2	225.6, 192.5	199.8, 208.8, 271.0
PEG	196.0, 272.6	198.2, 278.4	288.2, 235.0, 202.6

Polyethylene glycol (20%) showed an effect on the λmax in the case of benign and malignant postmenopausal tumors. The λmax at (227.6 and 226.2nm) disappeared at pH 8.2 and a new λmax at (272.6 and 278.4 nm) were obtained, while λmax at (192.6 and 194.0 nm) were both increased in the presence of PEG. The new λmax at 272.6 and 278.4 nm are both related to $\pi\rightarrow\pi^*$ transition of the aromatic ring of tyrosyl residue. In the case of benign premenopausal tumor, the presence of PEG showed an effect on the λmax of tyrosyl residue at pH 7.4. There is a significant red shift in this band from (276.0 to 288.2 nm) due to the $n\rightarrow\pi^*$ transition which occur at longer wavelength in low polar solvent [229]. The λmax at 224.4 nm was increased to 235.0 nm and a new λmax at 202.6 nm was obtained which was assigned to the amide group in polypeptide bond.

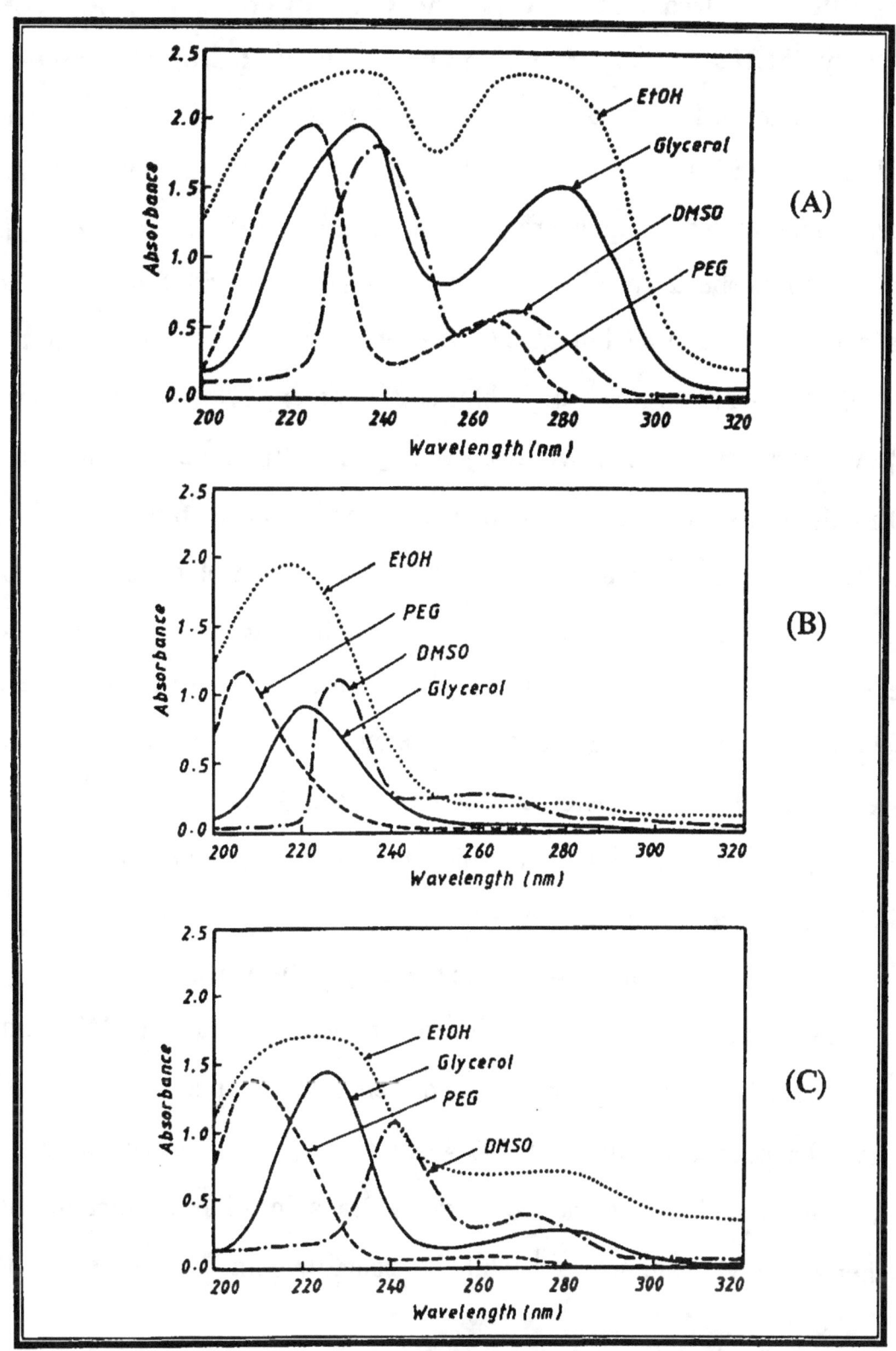

Figure (3-20): The effect of 20% (Ethanol, PEG, DMSO and Glycerol) on
         U.V spectrum of
              A-hLH,
              B-hLH antibody,
              C- (hLH-antibody) complex, at neutral pH.
         Details are described in section (2.3.1.2).

2- The Effect of Urea and KCl on The U.V. Spectrum of hLH, hLH Antibody, (hLH-Antibody) Complex and (^{125}I-hLH-Receptor) Complex

The effect of 8M urea, 0.03M KCl and a mixture of 1:1 of 8M urea and 0.03M KCl on the U.V spectrum at neutral pH were examined in this experiment. The values of λmax are illustrated in table (3-9), figure (3-21 A, B & C) shows these spectrum. When comparing these values with those obtained in the absence of urea or KCl in table (3-7 A) at neutral pH, it seems that in the presence of 8M urea there was a slight blue shift in λmax$_1$ 278.0 nm to 277.0nm in the U.V spectrum of hLH, while at λmax$_2$ 228.6 nm there was a red shift (~7nm) in this λmax of the polypeptide bond with contribution of histidyl residue. The blue shift is due to that the protein is unfolded in the high concentration of urea, the chromophores buried in the interior are transformed into the solvent. This transfer produce a blue shift in the absorption of these chromophores (tyrosyl), giving rise to a moderate decrease in the absorption at this wavelength. Also this transfer leads to a new λmax to be appear [222]. While the shift in λmax$_2$ is due to the intermolecular hydrogen bonding between the oxygen of the amide group and the solvent.

A red shift is also noticed (the same table) in the λmax$_1$ at 233.4 nm of hLH antibody spectrum (which is assigned to polypeptide bond) to 240.8 nm, while λmax$_2$ at 204.6 nm disappeared in the presence of 8M urea.

A new λmax was obtained at 247.4 nm in the (hLH-antibody) complex spectrum which could be assigned to the $\pi \rightarrow \pi^*$ transitions in the aromatic ring of the phenyl alanyl residues [229], while λmax at 207.0 nm disappeared in the presence of 8 M urea.

When 0.03M KCl was used, there were no alterations in the positions of the λmax$_1$ of the tyrosyl at pH (7.3) in hLH spectrum, and there was a significant red shift (~13 nm) in the λmax$_2$, while λmax$_1$ at 233.4 nm in hLH antibody spectrum disappeared in the presence of 0.03M KCl and λmax$_2$ increased from

204.6 to 211.4 nm. Such a red shift can arise by introducing positive (K^+)or negative (Cl^-) charges near the chromophore (the amide group) which might interact directly with the π electron system of the amide group [230].

Two new peaks appeared in (hLH-antibody) complex spectrum when 0.03M KCl was used. the first peak at 277.2 nm was assigned to tyrosine and the second peak at 235.0 nm was assigned to the polypeptide bond, while λmax at 207.0 nm of the amide group was disappeared.

Table (3-9): **The effect of urea and KCl on the λmax of hLH, hLH antibody and (hLH-Antibody) complex U.V. spectrum at neutral pH.**
Details are described in section (2.2.3.2)

Solvent	hLH λmax(nm)	hLH antibody λmax(nm)	(hLH-antibody) complex λmax(nm)
8M urea	277.0, 235.8	240.8	247.4
0.03M KCl	278.2, 242.2	211.4	277.8, 220.6
8M urea+0.03 M KCl (1:1)	278.0, 240.0	232.4	277.8, 235.0

When 8M urea was mixed with 0.03M KCl, there was a slight blue shift in λmax₁ of hLH spectrum and a red shift was obtained in λmax₂ from 228.6 to 240.6 nm in the same spectrum. A slight blue shift was noticed in λmax₁ of hLH antibody spectrum and disappearance of λmax₂ in the same spectrum, while a new λmax at 277.8 and 235.0 nm were appeared in (hLH-antibody) complex spectrum and λmax at 207.0nm was disappeared. The same peak at 277.8nm appeared when 0.03M KCl only was used, this means that this peak was appeared due to the effect of KCl but not to urea.

As was seen, the changes in absorption near 230nm were larger than those near 278 nm. This was noted by *Glazer* who reported that solvent perturbation or denaturation of protein produce a much larger changes in absorption near 230nm than 280 nm. Some of this changes in absorption may be produced by changes in the n→π* absorption of the polypeptide bonds in the protein either because of a change in their geometrical arrangement, or because of an environmental changes [222].

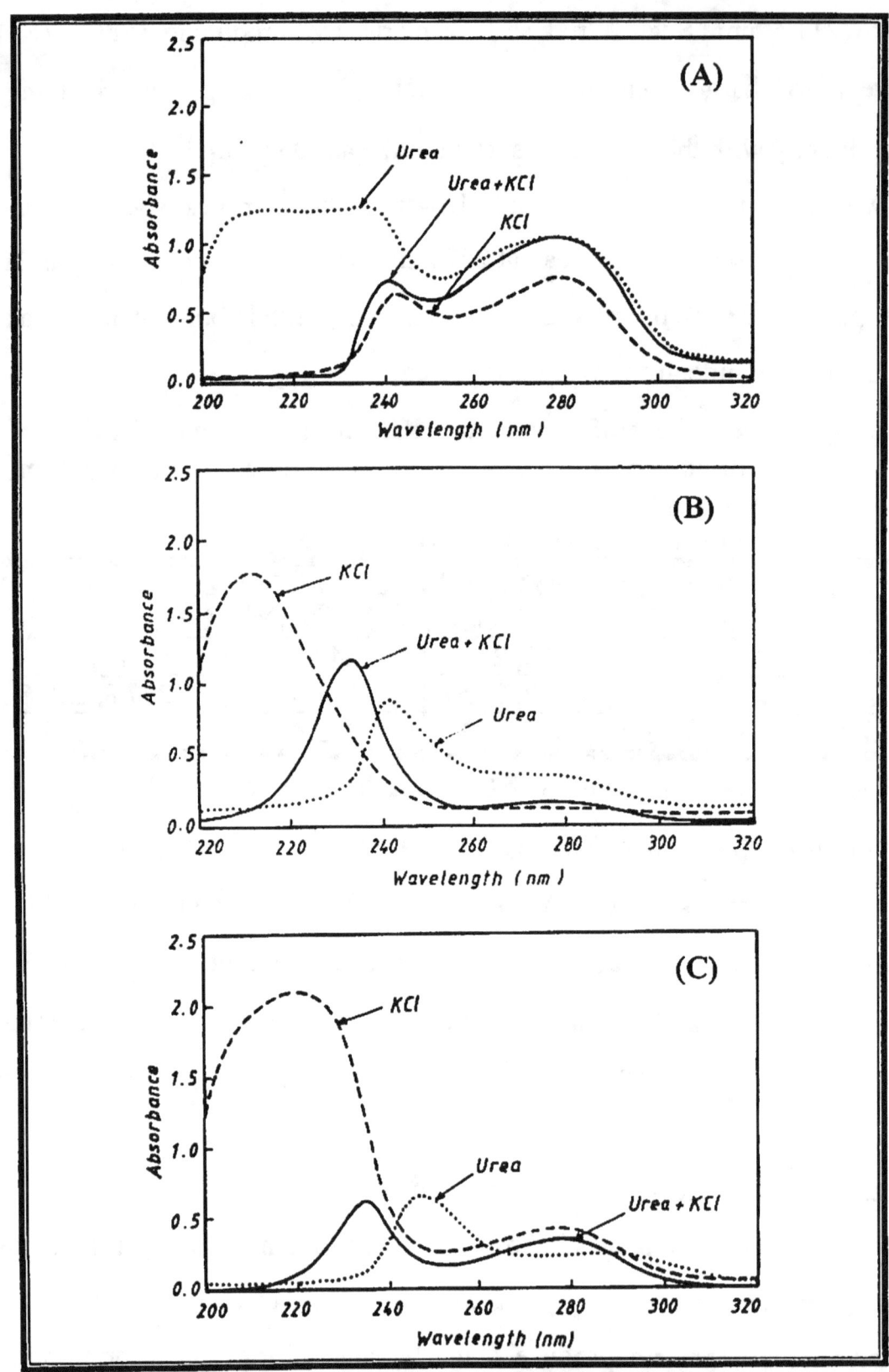

Figure (3-21):   The effect of (8M) urea, 0.03M KCl and 8M urea + 0.03M KCl (1:1) on U.V spectrum of

A-hLH,

B-hLH antibody,

C- (hLH-antibody) complex, at neutral pH.

Details are described in section (2.3.1.2).

The effect of 8M urea only was studied in the cases of benign and malignant ovarian tumor. Table (3-10) shows the λmax values obtained from this experiment. It seemed that there is a slight blue shift in λmax from (192.6 and 194.0 nm to 191.2 and 192.6 nm) and a red shift to a longer wavelength in λmax (227.6 and 226.2 nm to 233.8 and 235.2 nm) and also two new absorption peaks were appeared at (265.2, 258.8 nm) and (277.4, 278.0 nm) in the case of benign and malignant postmenopausal ovarian tumors respectively. This new peak at (265.2, 258.8 nm) was assigned to phenyl alanine while the second peak at λmax (277.4, 278.0 nm) was assigned to tyrosine. while in the case of benign premenopausal ovarian tumor, the (^{125}I-hLH-receptor) complex molecules were highly changed. Hence, an increased number of chromophores came to surface in the presence of 8M urea. the peak at 192.4 nm is assigned to tyrosine; the peak at 249.6 nm is assigned to phenyl alanine and the peak at 208.2 nm is assigned to the amide group in the polypeptide bond and also a slight blue shift in λmax$_1$ (2 nm), while λmax$_2$ shifted toward a longer wavelength (λmax 236.4nm). These results indicate that urea affects (^{125}I-hLH-receptor) complex molecule structurally since many chromophores which were embedded in an interior region of the protein where they were inaccessible to the solvent came into contact with it due to the unfolding of the molecule. Hence, different spectra were obtained [230].

Table (3-10): The effects of (8M) urea on the λmax of (^{125}I-hLH- receptor) complex U.V spectrum.
Details are described in section (2.3.1.2)

Group	pH	Without 8M urea λmax (nm)	With 8M urea λmax (nm)
premenopausal patients with benign tumor	7.4	276.0, 224.0	274.2, 249.6, 236.4, 208.2, 192.4
postmenopausal patients with benign tumor	8.2	227.6, 192.6	277.4, 265.2, 233.8, 191.2
postmenopausal patients with malignant tumor	8.2	226.2, 194.0	278.0, 258.8, 235.2, 192.6

3.4.2.3 Observation of the Helix-Coil Transition of hLH Antibody (Denaturation)

The effect of different concentrations of NaCl on the thermal stability of hLH antibody molecule was examined in this experiment. The values of absorbance at λmax (292 and 295 nm) for tryptophyl and tyrosyl residues respectively, in two different concentrations of NaCl 0.01M and 0.1M in a mixture of (20% ethylene glycol + 80% H_2O) are shown in figure (3-22).

As shown in figure (3-22), the absorbance of both tyrosine and typtophan reached higher absorbance at 50°C, in the presence of 0.01M NaCl and at 60°C, in the presence of 0.1M NaCl, although the absorbance at 0.1M NaCl was significantly lower than that in 0.01M NaCl, therefore higher concentration of NaCl causes more stabilization for hLH antibody molecule.

The increment in the absorbance of both tyrosyl and tryptophyl residues with increasing temperature could be due to that buried chromophores became exposed to the solvent during thermal denaturation, thus a helix coil transition for hLH antibody molecule was observed [222].

The decrease of absorbance in presence of 0.1 M NaCl as compared with that in 0.01 M NaCl could be due to salt concentration. Each protein in solution containing salts will collect about it a counter ion atmosphere enriched in appositely charged small ion, (chloride ion and sodium ion), and such a cloud of ions will tend to screen the protein, the larger the concentration of small ion present, the more effective this electrostatic screening will be, and decrement in the absorption intensity will be observed [233].

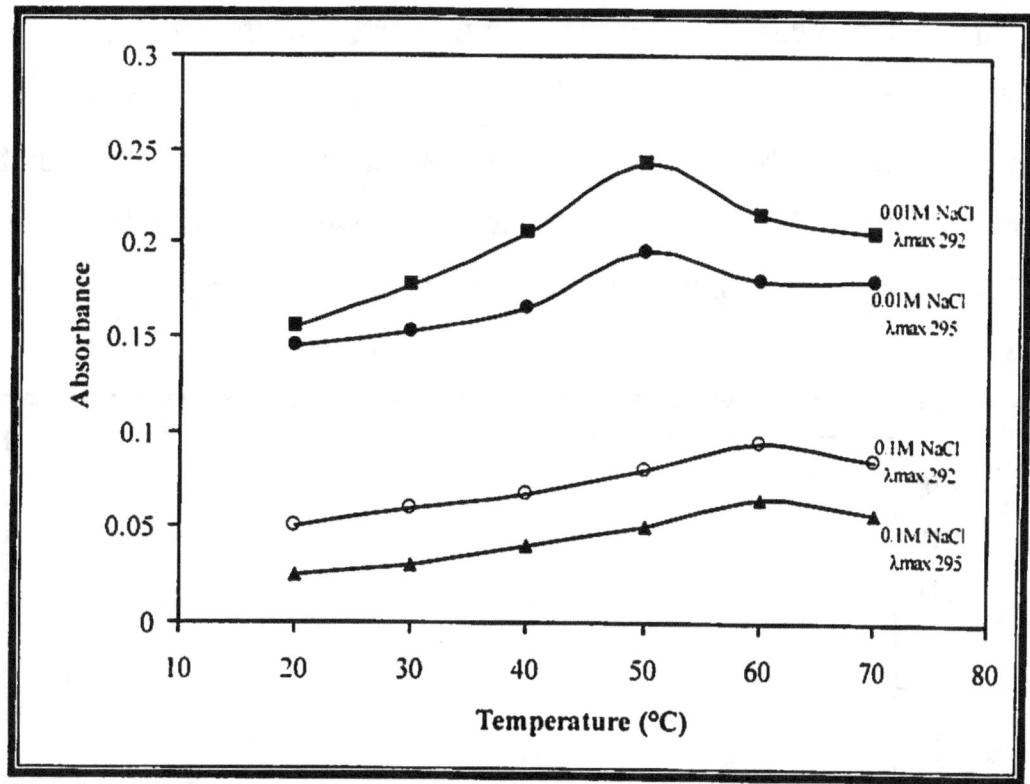

Figure (3-22): The effect of NaCl concentration on thermal stability of hLH antibody molecule.
Details are described in section (2.3.1.3)

3.4.2.4 Spectrophotometric Titration of hLH Antibody

Spectrophotometric titration is the following of the change in absorbance of the chromophore with increasing pH[224]. Many studies of protein structure require the determination of pK values for proton dissociation from ionizable amino acid side chains, because these values give an indication of the location of the amino acid in the protein. This can often be done spectrophotometrically because dissociation often changes the spectrum of one of the chromophores, the observation of tyrosine dissociation was performed by measuring the absorption at 295 nm (λmax for the ionized form of tyrosine), and the observation of histidine dissociation was carried out by measuring the absorption at 211 nm.

Figure (3-22 A & B) shows the pH titration curve of hLH antibody for histidine and tyrosine. Curve (B) shows that the pKa for Histidine is (5.6), while the pKa for tyrosine is (11.5). From the same figure, it is concluded that about

50% of histidine residues are located on the surface of the protein, while the other 50% are buried interior the hLH antibody molecule. While from curve B, it seems that about 60% of tyrosyl residues are internal and in a strongly non polar environment and therefore not titratable. The two curves also illustrate the low content of tyrosine compared to the content of histidine in hLH antibody molecule.

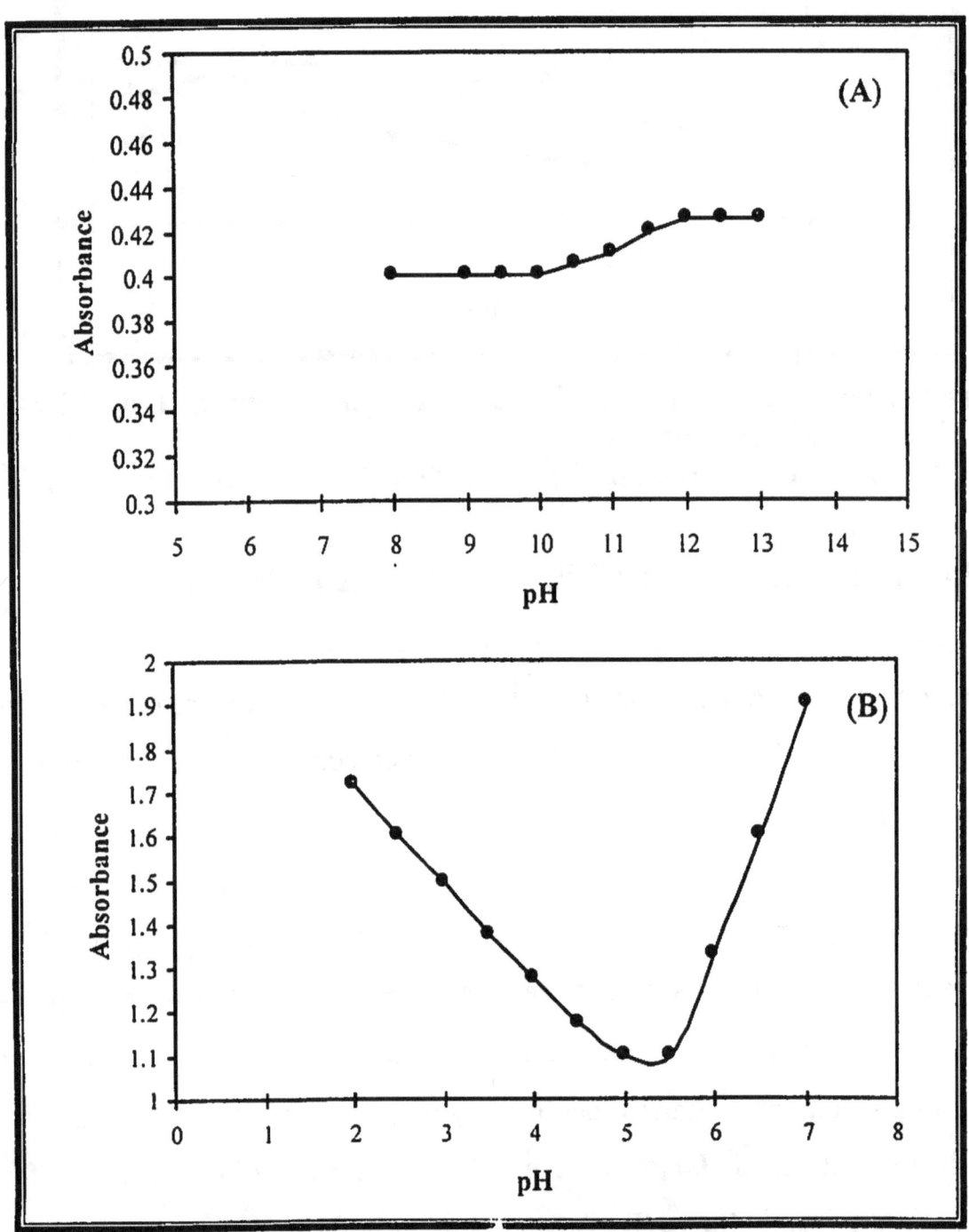

Figure (3-23): Spectrophotometric titration of hLH-antibody for:
(A) Tyrosine   (B) Histidine
Details are described in section (2.3.1.4)

Conclusions

1. A higher incidence of luteinizing hormone receptors was obtained in malignant than in benign serous ovarian tumors. Also a higher incidence of LH receptors was obtained in postmenopausal patients with benign serous ovarian tumors than in premenopausal patients with benign serous tumors.

2. The developed protocol for the assay of LH receptors is capable to analyze these receptors and the procedure is suitable for the assessment of LH receptors in benign and malignant serous ovarian tumors.

3. The Kinetic studies of the ^{125}I-hLH binding to its receptors in pre- and postmenopausal patients with serous ovarian tumors showed that the binding reaction is pH, time and temperature dependent process. The results, indicate that the reaction fits pseudo-first order kinetic.

4. The results obtained from the thermodynamic studies on the association of LH with its receptors, indicate that the binding reaction was entropically driven ($\Delta S° > 0$).

5. A characteristic spectrum was obtained from the spectroscopic studies on hLH, hLH antibody, (hLH-antibody) complex and (^{125}I-hLH-receptor) complex of pre- and postmenopausal serous ovarian tumors. This indicated that different compounds were obtained from binding reaction.

Future Work

1- Purification of the LH receptors in different groups of ovarian tumors.

2- Molecular characterization of these receptors of ovarian tumors.

3- Molecular biology of LH receptors in ovarian tumors.

4- Correlation of these receptors of ovarian tumors with those of other specific tumor markers.

5- Spectroscopic studies on purified LH receptor in ovarian tumors.

References

1. Arthur C. Guyton, John E. Hall. *Text Book of Medical Physiology*. 9th. ed. W.B. Saunders company. 1996, pp. 1017-1022.

2. Juan Rosai. *Ackerman's: Surgical Pathology*. 8th. ed. Mosby - year book. 1996, pp. 1461-1492.

3. Carol Mattson Porth. *Pathophysiology*. 4th. ed. J.B. Lippincott company. Philadelphia. 1994, p. 749.

4. Ivan Damjanov, James Linder. *Anderson's Pathology*. 10th. ed. Mosby - year book. 1996, p. 2278.

5. William Jubiz. *Endocrinology: A Logical Approach For Clinicians*. 2nd. ed. McGraw-Hill book company. 1987, p. 436.

6. Rod R. Seeley, Trent D. Stephens, Philip Tate. *Essentials of Anatomy And Physiology*. 4th. ed. McGraw-Hill book company. 1998, pp. 528, 935-942.

7. Clement PB. *Ovary: Histology For Pathologists*. New York, Raven Press. 1992. p.758.

8. Bruce R. Carr, Richard E. Blackwell. *Text Book Of Reproductive Medicine*. 2nd. ed. Appleton and lange. 1998, pp. 213-225.

9. Gerard J. Tortora, Sandra Reynold Grabowsk. *Principles Of Anatomy And Physiology*. 9th. ed. John Wiley and Sons Inc. New York. 2000, pp. 568-572.

10. Clement PB. *Am. J. Surg. Pathol*. 1987; 11:277.

11. Geoffrey Chamberlain. *Gynecology By Ten Teachers*. 16th. ed. Oxford University Press, New York. 1995, p. 142.

12. Jan Forth, D.N., Scott, J.R. *Obstetrics and Gynaecology*. 5th. ed. J.B. Lippincott company. Philadelphia. 1986, p. 918.

13. Adashi EY. *Hum. Reprod*. 1994; 9:815.

14. Upadhyay S., Zamboni L., Ectopi C. *Proc. Natal. Acad. Sci. USA*. 1982; 79:6584.

15. Dawson AB, McCabe M. *J. Morphol*. 1951; 88: 543.

16. Robert M. Berne, Matthew N. Levy. *Physiology*. 3rd. ed. Mosby-year book. 1993, p. 1003.

17. Carr B.R. *Disorder Of The Ovary And Female Reproductive Tract*. 8th. ed. Philadelphia, W.B. Saunders company. 1992, pp. 733-798.

18. Van Nagell, J.R., Powell, D.F., and Gay E.C. *Female Patients*. 1983; 8: 24.

19. Iemarie WJ., Conly PW, Moffett A. *Am. J. Obstet. Gynecol*. 1971; 110:612.

20. Strauss J.F., Gurpide E. *The Endometrium: Regulation And Dysfunction In Reproductive Endocrinology*. Philadelphia, W.B. Saunders company. 1992, pp. 309-356.

21. William W. Hay, Jr. Anthony R. Hayward, Myron J. Ievin, Judith M. Sondheirmer. *Current: Pediatric Diagnosis And Treatment*. 14th. ed. Appleton and Lang company. 1999, p. 604.

22. Lipsett M., Jaffe RB. *Steriod Hormones*, 2nd. ed. Philadelphia W.B. Saunders. 1986, pp. 140-153.

23. Richard C. Tilton, Albert Balows, David C. Hohnadel, Robert F. Reiss. *Clinical Laboratory Medicine*. Mosby-year book. 1992, p. 285.

24. J-de Brux, J-p Gautray. *Clinical Pathology Of The Endocrine Ovary*. Philadelphia. 1984, pp. 115-118.

25. Brenner P.F. and Mishell D.R. *Menopause: Infertility, Contraception And Reproductive Endocrinology*. 3rd. ed. Blakwell Scientific Publications. 1991, pp. 214-253.

26. Edward E. Partridge, Jerri Linn Phillips, Herman R. Menck. *Cancer*. 1996; 78: 2236.

27. Sauramo H., *Ann. Clin. Gynacol. Fenn*. 1952; 4:1.

28. Nagamani M., Stuart CA. *J. Clin. Endocrinol. Metab*. 1992; 74:172.

29. Thomas J. Nowak, A. Gordon Handford. *Essentials Of Pathophysiology*. 2nd.ed. WCB. McGraw-Hill. 1999, pp. 501-504.

30. Eric V. Mackay, Norman A. Beischer, Lloyd W. Cox, Carl Wood. *Illustrated Text Beek Of Gynaecology*. W.B. Saunders company. 1983, p. 282.

31. Charles R. Whitefield. *Dewhurst's Text Book Of Obstetrics And Gynaecology*. 5th. ed. Black well science. 1995, pp. 759-774.

32. Anthonys Fauci, Eugene Braunwald, Kury J. Isselbacher, Jean D. Wilson, Joseph B. Martin, Dennis L. Kasper, Stephen L. Hauser, Dan L. Longo. *Harrison's: principles of internal medicine*. 14th. ed. McGraw-Hill company. 1998, pp. 605-608.

33. Herbert A. Fritsche, Robert C. Bast. *Clinical Chemistry* 1998; 4: 1379.

34. Berek JS, Hacker NF. *Non Epithelial Ovarian And Tubal Cancer: Practical Gynecologic Oncology*. 2nd. ed. Williams and Wilkins company. 1995, p. 295.

35. Fuys Woodruff JD. *Pathology: Practical Gynecologic Oncology*. 2nd. ed. Williams and Wilkins company. 1995, p. 1079.

36. Rusell P. *Clin. Obstet. Gynaecol*. 1984; 11: 259.

37. Caroline Van Haaften, Cinda M. Boyer. *Cancer*. 1996; 77:1131,2092.

38. Karlan BY, Platt LD. *Gynecol. Oncol*. 1994; 55: S 28.

39. Ministry of Health. *Results of Iraqi Cancer Registry*, 1976-1997.

40. Watter J. Daly, J. Donald Eston, John J. Hutton, Peter O. Kohler, Robert A. O'Rourke, Merle A. Sande, Jay H. Stein, Jerry S. Trier, Nathan J. Zuaifler. *Internal Medicine*. 2nd. ed. Little, Brown and company, Boston. 1987, pp. 1121, 1811-1817.

41. American Cancer Society. *Cancer Facts And Figures*. New York. 1992, pp. 11-13.

42. Panndge E., Philip J., Menck H. *Cancer*. 1996; 78: 2236.

43. Henricksen R., Strang P., Vilander E. *Gynecol. Oncol*. 1994; 53: 301.

44. Harlow BL, Weiss NS, Roth GJ, Chu J, Daling JR. *Cancer Res*. 1988; 48: 5849.
45. Gross TP., Schlesselman JJ. *Obstet. Gynecol. Oncol*. 1994; 83:419.
46. Santoso JT, Tang D. C, Lane SB. *Gynecol. Oncol*. 1995; 59: 171.
47. Nigro JM, Baker S., Preisinger Ac. *Nature*. 1989; 342: 705.
48. Hollstein M., Sidransky D., Vogelstein B. *Science*; 1991; 253: 49.
49. Seidman JD. *Obstet Gynecol*. 1993; 81: 643.
50. Mark JR, Davidoff AM, Kerns BJ. *Cancer Res*. 1991; 51: 2979.
51. Frank TS, Bartos RE, Haefner HK. *Mod. Pathol*. 1994; 7:3.
52. Bourne TH, Whitehead ML, Campbell S, *Gynecol. Oncol*. 1991; 43: 92.
53. Bell DA, Scully RE. *Cancer*. 1994; 73: 1859.
54. Daniel L. Clarke-Pearson, M. Yousif Dawood. *Green's Gynecology: Essentials Of Clinical Practice*. 4th. ed. Little, Brown and company. 1990, pp. 531-541.
55. Welch WR, Cassells S, Scully RE, *Am. J. Obstet Gynecol*. 1983; 147: 1.
56. Negri E., Franceschi S., Tzonou A. *Int. J. Cancer*. 1991; 49: 50.
57. Cramer DW, Welch WR. *J. Natl. Cancer Inst*. 1983; 71: 717.
58. Goldberg GL, Runowicz CD. *Am. J. Obstet Gynecol*. 1992; 166: 853.
59. James F. Holland, Robert C. Bast, Donald L., Emill Frei, Donald W. Kufe, Ralph R. Weichselbaum. *Cancer Medicine*. 4th. ed. Williams and Wilkins awaverly company. 1997, p. 284.
60. Fathalla MF. *Lancet*. 1971; 2: 163.
61. Vladimir Bychkov, John H. Isaacs. *Pathology In The Practice Of Gynecology*. Mosby-year book. 1995, pp. 274-279.
62. Hart WR. Hum. *Pathol*. 1977; 8: 541.
63. Julian GC, Woodruff JD. *Obstet. Gynecol*. 1972; 40: 860.
64. Vincent T. Devita, Samuel Hellman, Steven A. *Cancer: Principles and Practice Of Oncology*. 5th. ed. Lippincott-Raven. 1997, pp. 1504-1507.
65. Salls, Stone Ml .*Prog. Clin. Cancer* 1973 ; 5: 249 .
66. Yancik R . *Cancer*. 1993 ;74: 1995 .
67. Schwartz PE, Taylor Kjw . *Ovarian cancer : Epidemiological Perspectives With Developments In Early Diagnosis*. The Parthenon publishing group. New York .1994, p. 257.
68. James B. Wyngaarden Lloyd H .Smith , Jr. , J. claude bennett . *Cecil : Text Book Of Medicine* . 20th .ed . W.B .Saunders company .1996 ,pp1021-1022.
69. Ozols RF.*Gynecol Oncol* . 1994 : 55: S168 .
70. Finkler NJ, Kopnick SJ, Griffiths CT, Knapp RG. *Gynecol. Oncol*. 1988; 29: 356.
71. Lavin PT, Knapp RG, Malkasian C. *Obstet Gynecol*. 1987; 69: 22.
72. Bast RC, Kluy TL, ST John E, Jenison E, Niloff JM, Lazarus H. *N. Engl. J. Med*. 1983; 309: 169.
73. Nustad K, Bast RC, Obrien T., Nilzson O., Seguin P., Suresh M. *Tumor Biol*. 1996; 17: 176.

74. Jacobs I., Bast RC. *Hum. Reprod.* 1989; 4:1.
75. Borne T., Campbell S., Steer C., Whitehead MI. and Collins W. *BMJ.* 1989; 299: 1367.
76. Milner BJ., Allan LA., Eccles DM. *Cancer Res.* 1993; 53: 2128.
77. Kiyokawa T. *Int. J. Gynecol. Pathol.* 1994; 13: 311.
78. Vincent T., De Vita Jr., Samuel Hellman and Steven A. *Important Advances In Oncology.* Philadelphia, Lippincott-Raven. 1996, pp. 37-39.
79. Lynn C. Hartmann, karl C. Podratz, Garyl Keeney, Nermeen A. Kamel, John H. Edmonson, Joseph P. Grill, John Q. Su, Jerry A. Katzmann, and Patrick C. Roke. *J. Clin. Oncol.* 1994; 12: 64.
80. Yabushita H., Masuda T., Ogawa A., Noguchi M. and Ishihak M. *Gynecol. Oncol.* 1988; 29: 66.
81. Soper JT., Hunter VJ., Daly L., Tanner M., Creasman W. and Bast RC. *Obstet Gynecol.* 1990; 75: 249.
82. Thor A., Gorstein F., Ohuchi N., Szpak C., Johnston W., Schlom J. *J. Natl. Cancer.* Inst. 1986; 76: 955.
83. Einhorn N, Knapp R C, Bast R. C.Jr., Zurawski VR. Jr. *Acta Oncol*, 1989; 28:655.
84. Frederick P. Zuspan, Steven G.Gabbe,Thomas J. Garite, Moon H. Kim, Alberto Manetta. *Am. J. Obstet. Gynecol*,1994; 171:1183.
85. Bostwick DG., Tazelaar HD., Ballon SC. Hendrickson MR. and Kempson RL. *Cancer.* 1986; 58: 2052.
86. Tazelaar HD., Bostwick DG, Ballon SC., Henrickson MR. and Kempson RL. *Obstet Gynecol.* 1985; 66: 417.
87. McGuire WP. *Cancer* 1993; 71: 1514.
88. Alexander W. Kennedy and William R. Hart. *Cancer.* 1996; 78: 278.
89. Thigpen JT, Lambuth BW. and Vance RB., *Semin. Oncol.* 1991; 18: 596.
90. Cannistra SA. *N. Engl. J. Med.* 1993; 329: 1550.
91. Hacker NF, Berek JS, Lagasse LD., Nietberg RK. and Elashoff RM. *Obstet Gynecol.* 1983; 61: 413.
92. Vergote IB., Vergote De Vos LN. and Abeler VM. *Cancer.* 1992; 69: 741.
93. Rosenshein NB. *Clin. Obstet Gynecol.* 1983; 10: 279.
94. Neijt JP., Ten Bokkel Huinink WW., Vander Burg M., Vanoosteron AT, Willemse PHB and Heintz APM. *J. Clin. Oncol.* 1987; 5: 1157.
95. McGuire WP., Hoskins WJ, Brady MF, Kucera PR, Partvidge EE. and Look KY., *N. Engl. J. Med.* 1996; 334:1.
96. Kohler Dr. and Gdldspiel Br. *Pharmaco. Therapy.* 1994; 14: 3.
97. Ozols RF. *Ann. Med.* 1995; 27: 127.
98. Allan Covens, Sylvain Boucher, Kathie Roche, Moira Macdonald, Daneil Pettitt, Bruno Jolain, Eric Souetre and Marc Riviere. *Cancer* 1996; 77: 2086.
99. L. Migliette, D. Amoroso and M. Bruzzona. *Oncology.* 1997; 54: 102.
100. Friedman JB.and Weiss NS. N. *Engl. J. Med.* 1990; 322: 1079.

101. Miller DS., Spirtos NM., Ballon SC., Cox RS., Seriero OM. and Teng NN. *Cancer* 1992; 69: 502.

102. Seifer DB, Seriero OM. and Miller DS. *Cancer* 1986; 57: 530.

103. Podratz KC, Mal. Kasiam GD., Hilton JF, Harris EA. and Gaffey TA., *Am. J. Obstet Gynecol.* 1985; 152: 230.

104. Smirz LR, Stechman FB., Ulbright TM., Sutton GP. and Ehrlich CE. *Am. J. Obstet Gynecol.* 1985; 152:661.

105. M. Vassilomanolakis, G Koumakis, V. Barbounis, H. Hajichriston, ST. Sousis and A Efremidis. *Oncology.* 1997; 54: 199.

106. Janne O., Kauppila A., Syrjala P.and Vihko R. *Int. J. Cancer.* 1980; 25: 175.

107. Ford LL., Berek JS., Lagasse LD., Hacker NF., Heins Y., Esmailian F., Leuchter RS. and Delang RJ. *Gynecol. Oncol.* 1983; 15: 299.

108. Kuhnel R., De Graaf J., Rao Br. and Stolk JC. *J. Steroid Biochem.* 1987; 26: 393.

109. Slotman B.J. Rao Br. *Cancer J.* 1989; 11: 373.

110. Malkasiam GD, Melton LJ., Brien PC. and Greene MH., *Am. J. Obstet Gynecol.* 1984; 149: 274.

111. Beller U., Bigelow B., Beckman EM., Brown B.and Demopoulos RI. *Gynecol Oncol.* 1983; 15: 422.

112. Thigpen T., Brady MF, Omura GA., Creasman WT, McGuire WP., Hoskins WJ.and Williams S. *Cancer.* 1993; 71: 606.

113. Demopoulos RI., Bigelow B., Blaustein A., Chait J., Gutman E.and Dubin N. *Obstet. Gynecol.* 1984; 64: 557.

114. Nikrui N. *Gynecol Pathol.* 1981; 12: 107.

115. Bostwick DG, Tazelaar HD., Ballon SC, Hendrickson MR. and Kempson RL. *Cancer.* 1986; 58: 2052.

116. Kurman R.J. and Trimble CL. *Int. Gynecol. Pathol.* 1993; 12: 120.

117. Sorbe B.and Fankendal Bo. *Obstet. Gynecol.* 1982; 59: 576.

118. Swenerton Kd., Hislop TG., Spinelli J., Leriche JC. Yang N. and Boyes DA. *Obstet Gynecol.* 1985; 62: 264.

119. Tornos C., Silva EG, Khorana SM and Burke TW. *Am. J. Surg. Pathol.* 1994; 18: 687.

120. Sorb B. and Frankendal Bo. *Gynecol. Oncol.* 1982; 14: 6.

121. Sevelda P., Dittrich C. and Salzer H. *Gynecol. Oncol.*, 1989; 35: 321.

122. Erhardt K., Auer G., Bjorkholm E., Forsslund G., Moberger B., Silfversward C., Wicksell G. and Zetterberg A. *Am. J. Obstet. Gynecol.* 1985; 151: 356.

123. Rodenburg CJ., Cornelisse CJ., Heintz PAM., Hermans J. and Fleuren GJ. *Cancer* 1987; 59: 317.

124. Marker APH., Kristensen GB., Kaem J., Bomer OP., Abeler VM. and Trope CG. *Obstet. Gynecol.* 1992; 79: 1002.

125. Daniela Massi, Tommaso Susini, Luciano Savino, Vieri Boddi, Gianni Amunni, Maurizio Colafranceschi. *Cancer* 1996; 77: 1131.

126. Brioschi PA, Bischof P, Rapin C, De Roten M, Iriono, Krauer F. *Gynecol Oncol*, 1985; 21:1.

127. Schwartz PE, Chambers SK, Chambers JT, Gutmann J, Katopodis N, Foemmal R. *Cancer*, 1987; 60:353.

128. Pierce, J.C.. *Obstet. Gynecol. News*. 1992; 27:1.

129. Jacobs I, Bridges J., Reynolds C, Stabile I, Kemsley P., Grudzin S. and Kas J. *Lancet* 1988; 2: 268.

130. Pierce, J.C. and Parsons, T. F. *Annal. Rev. Biochem*. 1981; 50: 465.

131. Dennis Schulster and Alexander Levitzki. *Cellular Receptors For Hormones And Neurotransmitters*.John Wiley and sons.1980. pp.149-159.

132. Robert H. Williams, *Text Book Of Endocrinology* 5th. ed., W.B. Saunders company, London. 1974, p. 40.

133. Hockscher SR., Sairam MR and Ascolim. *Endocrinology*. 1991; 128: 2837.

134. Baenziger Ju, Kumars, Brodbeck RM, Smith RL and Beranek MC. *Proceedings Of The National Academy Of Sciences Of The USA*. 1992; 89: 334.

135. Willis, M. *Biol. Rv*. 1975; 50: 35.

136. Talmadge K., Vamvako Poulos, N.C. and Fiddes J.C. *Nature*. 1984; 307:37.

137. Sairam, M.R. and Li, C.H. *Hormonal Proteins And Peptides*, vol. 6. Academic Press, New York. 1978, pp. 1-56.

138. M. Wallis, S.L. Howell and K.W. Taylor. *The Biochemistry Of The Polypeptide Hormones*. John Wiley and Sons Inc., 1985, pp. 29-36, 415-428.

139. Roberk K. Murray, Daryl K. Granner, Peter A. Mayes and Victor W. Rodwell. *Harper's Biochemistry*, 25th. ed. Appleton and Lange. 2000, pp. 500-550.

140. Chatterjee M. and Munro H.N. *Vitamins and Hormones*. 1977; 35: 149.

141. Fiddes J.C. and Goodman H.M. *Nature*. 1980; 286: 684.

142. Shome B. and Parlow A.F. *J. Clin. Endocrinol. Metab*. 1974; 39: 199.

143. Cornell J.S. and Pierce J.G. *J. Biol. Chem*. 1973; 248: 4327.

144. Walter Hubl. *Hormone-Diagnosis For Fertility Disorders: Institute For Clinical Chemistry*. Dresden - Friedrichstadt Hospital, 1992, pp. 7, 52.

145. Tietz N., Ed. *Clinical Guide To Laboratory Tests*, 3rd. ed. W.B. Saunders company. Philadelphia. 1995.

146. Orten JM., Neuhaus OW. *Human Biochemistry*, 10th. ed., St. Louis, Mosby-year book. 1982, p. 601.

147. Emil L. Smith, Robert L. Hill, I. Robert Lehman, Robert J. Lefkowitz, Philip Hanler, Abraham White. *Principles Of Biochemistry: Mammalian Biochemistry*. 7th. ed. McGraw-Hill book company. 1985. pp. 376-378.

148. Carl A. Burtis, Edward R. Ashwood, *Tietz Text Book Of Clinical Chemistry* 3rd. ed., W.B. Saunders company. 1999. p. 1466.

149. GW Montgomery, ML Tate, HM Henry, JM Penty and RM Rohan. *J. Of Molecular Endocrinol.* 1995; 15: 259.

150. Combarnous. *Endocrine Review.* 1992; 13: 670.

151. Segaloff DL and Ascoli M. *Oxford Reviews Of Reproductive Biology*, 1992; 14: 141.

152. Monetta A., Pinto J.L., Larson J.E. *Obstetrics and Gynecology.* 1988; 72: 77.

153. Conn PM, Crowley WF. Jr. *N. Engl. J. Med.* 1991; 342: 93.

154. Knobil E. *Recent Prog. Hormone. Res.* 1980; 36: 53.

155. C.R.W. Edwards, I.A.D. Bouchier and C. Haslett. *Davidson's: Principles And Practice Of Medicine.* 17th. ed., Churchill livingstone. 1995, p. 672.

156. Knobil E., Plant TM and Wildt TL *Science*; 1980; 207: 1371.

157. Piva F., Motta M., and Matini L. *Endocrinology*, Gruneanel Starttton, New York. 1979, pp. 21-33.

158. Licht P., Popkoff H. and Farmer S. *Horm. Res.* 1977; 33: 169.

159. Wide L., *Hormone Assays And Their Clinical Application*, 4th. ed., Churchill liningstone, Edinburgh, 1976, pp. 87-140.

160. Roth J. *Methods In Enzymology.* 1975; 37: 66.

161. Odell W.D., and Daughaday W.H. *Principles Of Competitive Protein Binding Assay.* Lippincott company. Philadelphia. 1971. p. 543.

162. Yalow R.S. and Berson S.A., *J. Clin. Inves.* 1960; 39: 1157.

163. Alan H. Gowenlock, Janet R. McMurray, Donald M. Mclauchlan. *Varley's Practical Clinical Biochemistry.* 6th. ed. Heinemann medical books, London. 1988, pp. 110-114.

164. Hales C.N. and Randle P.J. *Biochem.J.* 1963; 88: 137.

165. Morgan C.R. and Lazarow A. *Diabetes.* 1963; 12: 115.

166. Beastall G., Ferguson K. and O'Reilly D. *Ann. Clin. Biochem*, 1987; 24: 246.

167. Hemmila I., Hottinen S. and Pettersson, K. *Clin. Chem.* 1987; 33: 2281.

168. Lowry O.H., Rosebrough N.J., Farr A.L. and Randall R.J. *J. Biol. Chem*; 1951; 193: 265.

169. Morris BJ. *Clinica Chimica* Acta. 1976, 73: 213.

170. Scopes R.K. *Protein Purification: Principles And Practice.* 2nd. ed; Springer Verlag. 1987; pp. 196-198.

171. Scatchard G. *Ann NY Acad Sci.* 1949, 51: 660.

172. Segal I.H. *Biochemical Calculations*, 2nd. ed. John Wiley and Sons Inc. 1976; p. 327.

173. Chamberlain J., Jargarinec N. and Ofner P., *Biochem. J.* 1966; 99:610.

174. Thompson S.A., Johnson M.P.and Brook S.C. *The prostate.* 1982; 3: 45.

175. CH.V. Rao and S. Mitra. *Biochimica Biophysica Acta.* 1979; 584: 454.

176. BlaaKaer J., Djursing H., Hording U., Bennett P., Toftager Larsen K., Bock JE. and Lebech PE. *Acta. Endocrinal. Copenh.* 1992; 127: 127.

177. Jeppsson RL. *Acta. Obstet. Gynecol. Scand.* 1986; 65: 207.

178. Kahn CR. *J. Cell. Biol.* 1976; 70: 261.

179. Fridovich, I. *Annu. Rev. Biochem.* 1975; 44: 147.

180. Joseph V. Princiotto and Edward J. Zapolsk. *Biochimica. Biophysica. Acta.* 1976; 428: 766.

181. Yehudith Amir-Zaltsman and Yoram Salomon. *Endocrinology* 1980; 106: 1166.

182. Hannu J. Raganiemi, Lars Ronnberg, Antti kauppila and Pekka Ylostalo. *J. Clin Endocrinol. Metab.* 1981;108:307.

183. Cameron J.L. and Richard L.Stouffer. *Endocrinology.* 1982; 110: 2059.

184. Jaquette J and Segaloff DL. *Endocrinology.* 1997; 138: 85.

185. Mrinalini C. RAO, Joanne S. Richards, A. Rees Midgley, J. Rand leo and E. Reichert. *Endocrinology.* 1977; 101: 512.

186. Richard L. Stouffer, Martin S. Grodin, John R. Davis and Earl A. Surwit. *J. Clin. Endocrinol. Metab.* 1984; 59: 441.

187. NaGuib A. Samaan, Kuo-Pao yang and Darrel N. Ward. *Endocrinology,* 1976; 98: 233.

188. Sandra Kammerman, Rita I. Demopoulos, Cynthia Raphael and Joel Ross. *Human Pathology.* 1981; 12: 886.

189. C.Y. Lee and R.J. Ryan. *Biochemistry.* 1973; 12: 4609.

190. Hannu J. Rajaniemi, A. Rees Midgley Jr., Joyce A. Duncan, and Leo E. Reichert. *Endocrinology.* 1977; 101: 898.

191. C.Y. Lee, C.B. Coulam, N.S. Jiang, and R.J. Ryan. *J. Clin. Endocrinol. Metab.* 1973; 36: 148.

192. Kolena J., Sebokova E., Arch Int. *Physiol. Biochem.* 1986; 94: 261.

193. Jia XC., Oikaw M., Bo M., Tanaka T., NY T., Boime I., Hsueh AJ. *Mol. Endocrinol.* 1991; 5: 759.

194. Joshi LR., Boland SR., Hewlett EL., and Katz MS. *Arch. Biochem. Biophys.* 1988; 261: 134.

195. Ralph H. Schwall and Gregory F. Erickson. *Endocrinology.* 1984; 114: 1114.

196. Sakai S., J. Diary. *Science.* 1994; 77: 433.

197. Melander W., Hovarth C., *Arch. Biochem. Biophys.* 1977; 183: 200.

198. Y. D. I. Chen and A.H. Payne. *Biochimica and Biophysica Res. comm.*

199. Leake A., Chrisholm G.D., Busuttil A. and Habib F.K., *Acta. Endocr. Copenh.* 1984; 105: 281.

200. Joan Reed M. and Stitch S.R. *J. Endocrin.* 1973; 58: 405.

201. V. Thambyra Jah, Salman Azhar and K.M.J. Menin. *Biochimica. Biophysica Acta.* 1976; 428: 35.

202. K.J. Ryan. *J. Clin. Endocrinol. Metab,* 1973; 36: 118.

203. Keinanen KP. and Rajaniemi HJ. *J. Clin. Endocrinol Metab.* 1988; 67: 228.

204. Yamoto M., Nakano R., Iwaski M., I koma H. and Furukawa K., *Obstet. Gynecol.* 1986; 68: 200.

205. Weiss N.S. Geneva. *International Union Against Cancer.* 1980.

206. Rajaniemi H., Kauppila A., Ronberg L., Selander K., Pystynen P. *Acta. Obstet. Gynecol. Scand* [Suppl]. 1981; 101: 83.

207. Vihco-Kk, Kujansuu-E, Morsky-P; Huhtaniemi-I, Punnonen-R. *Eur. J. Endocrinol.* 1996; 134:357.

208. Feng-Y; Zhang-X and Ge-B. *Chung-Hua-Fu-Chan-Ko-Tsa-Chih.* 1996; 31: 166.

209. Mandai-M, Konishi-I, Kuroda-H and Fukumoto-M. *Eur. J. Cancer.* 1997; 33:1501.

210. Cui-J, Guo-Y, You-Z. Chung-Hua-Fu-Chan-Ko-Tsa-Chih. 1997; 32:742.

211. Dufau, M. L., and K. J. Catt. *Gonadal Receptors For Luteinizing Hormone And Chorionic Gonadotropin.* Raven-press, New York, 1976, p. 135.

212. Haour F., and B. B. Saxena. *J Biol Chem.* 1974; 249: 2195.

213. Kananen-K and Markkula-M. *Mol. Endocrinol* 1995; 9:616.

214. Seeley D. H., Wang W. Y., Salhanick H. A.; *Biochem Biophys Acta*; 1980; 632: 536.

215. Weiland G. A. and Molinoff P. B. *Life Science*; 1981; 29: 314.

216. Nagasaw, H. *Prolactin In Lesion, In Breast, Uterous And Prostate.* Tokyo, Ckc press. 1989.

217. Nemethy G and Scherage HA. *J. Phys Cem.* 1962, 66: 1773.

218. Waelbroeck M., Van Obberghen E. and DeMeyts P. *J Biol Chem.* 1979, 254:7736.

219. Ross P.D. and subramanian S. *Biocem.* 1981; 20:3096.

220. Blumenthal D. K. and Stull J. T. *Biochem*; 1982; 21:2386.

221. Laporte D. C., Wierman B. M. and Storm D. R. *Biochem*; 1980; 19: 3814.

222. Leach, S. J. *Physical Principles And Techniques Of Protein Chemistry.* 1969, New York, Acadimic press, part A, pp. 102-170.

223. Mathews, Ch. K., Holde K. E. *Biochemistry.* California, The Benjamin/Cummings publishing Co. pp. 146-148.

224. Freifrlder, D. *Physical Biochemistry: Application to Biochemistry and Molecular Biology.* 2th. ed. 1982, San Francisco, W. H. Freeman and Company. pp. 494-512.

225. San. Y.; Bovey, D. A. *J. Biol. Chem.*; 1960; 235: 2818.

226. Nagacura S. and Baba H. *J. Am. Chem. Soc.* 1952; 74:5693.

227. Pimentel G. C. *J. Am Chem. Soc.*; 1957; 79:3323.

228. Brealy G. J. and Kaska M. *J. Am. Chem. Soc.*; 1955; 77:4462.

229. Silvestien, R. M.; Bassler, G. C.; Marril, T.C. *Spectrophotometric Identification Of Organic Compounds*. 1981, New York, John Wiley and sons. p.181.
230. Leach, S.J. and Scheray, H. A. *J. Biol. Chem.* 1960;10: 2827.

www.ingramcontent.com/pod-product-compliance
Lightning Source LLC
Chambersburg PA
CBHW080813180526
45168CB00006B/2429